4x4 Suspension Handbook

Handbook

TRENTON McGEE

S-A DESIGN

CarTech®

Edited by: Josh Brown
Designed by: Chris Fayers

ISBN 978-1-61325-082-2
Item SA137P

Printed in USA

Title Page:
Though it disappeared for a few years, Ford's radius arm/coil-spring suspension design continues today with the introduction of the late-model Super Duty. First released in 2005, coil-spring Super Dutys are top sellers. (Courtesy Superlift)

CarTech®

39966 Grand Avenue
North Branch, MN 55056
Telephone (651) 277-1200 • (800) 551-4754 • Fax: (651) 277-1203
www.cartechbooks.com

CONTENTS

PREFACE

This book is all about four-wheel-drive suspension systems and how they can be modified for better off-road performance. Unless you're a four-wheel drive nut (and there are plenty of you out there) you probably won't read this entire book. Instead, you might read the introduction and then skip ahead to the chapter that deals specifically with the type of vehicle you own. That's understandable, but I encourage you to fully read the first two chapters, as they deal with the basic theories and terminology common to all 4x4s, old and new. I also encourage you to scan the other chapters, as they contain lots of little kernels of information that can be applied to other types of vehicles. Also, having a basic understanding of other suspension types also allows you to be better informed about the subject as a whole.

What you are going to see in the pages that follow is an in-depth explanation of the most common four-wheel-drive suspension systems—how they work, why they work, and their good and bad qualities. Furthermore, I explain how these suspension systems are modified for better off-road performance. Ninety-nine and a half percent of the time that means lifting them, and many well-respected companies out there have been building high-quality lift systems for decades. That brings me to what you won't see in this book: The purpose is to provide a good knowledge base for the end consumer, not build a particular brand loyalty. The truth is that any suspension company that has been around for a while builds quality product; if not, lawsuits would have closed its doors long ago. In many cases, particularly with older vehicles, there is really only one way to lift a vehicle, and they all follow the same method. Other vehicles may have two or more schools of thought on modifying a particular suspen-sion design; in these cases I explain the differences along with their advantages and drawbacks as objectively as possible.

Another thing you are not going to see in this book is how to design and build your own lift system, or how to modify off-the-shelf systems for more lift. The suspension system of a vehicle is a highly complicated and critically important component of a vehicle, and modifying them safely is well beyond the capabilities of your average shade tree mechanic. If you find yourself thinking it can't be that hard, think again. I've seen a lot of butch stuff both on and off the road, and much more homebrew junk breaks out on the trail than off-the-shelf stuff. You wouldn't machine your own connecting rods and pistons for your engine, would you? Fabrication should be left to quali-fied professionals.

Now, kick back with your favorite beverage and read on.

SUSPENSION BASICS

Read This First!

While four-wheel-drive vehicles have been around in some shape or form since shortly after the invention of the automobile, we owe the four-wheel-drive systems as we know them today to the military. The need to transport men and materials to remote locations demanded vehicles with the capability of traversing muddy fields, ravines, hills, and rocky areas. By necessity, this meant all four wheels had to be able to put engine power on the ground. While many light, medium, and heavy-duty trucks were developed with this capability, most experts agree that it was the General Purpose Wagon, or GPW, that exposed the attributes of a 4x4 to the masses during World War II. Soldiers came to love the little ¼-ton GPW and nicknamed it the Jeep, a name that stuck so well it became the name for a legend that endures to this day.

But that's not where this story really begins. After the war there was a tremendous surplus of GPWs, and the military wisely decided to sell them off to the general public rather

Though the WWII-era Jeep may not have been the first 4x4 to be produced, it was the one that brought the capability of a four-wheel drive to the masses. Surplus GPWs and the civilian models that were offered immediately after World War II were inexpensive and allowed former GIs the ability to take their families places no sedan or station wagon could. It could be argued that the CJ-2A (the first civilian Jeep) and the models that followed have largely shaped the 4x4 market as we know it today. (Courtesy DaimlerChrysler)

than scrap them. Returning soldiers, flush with cash, could pick up a surplus Jeep that they knew and loved for next to nothing. Furthermore, Willys Overland, the patent holder for the Jeep's design, started churning out civilian versions (the CJ-2A) that were virtually identical to the military Jeeps, and these too were cheap. It wasn't long before folks started tinkering with these Jeeps to individualize them and better suit their needs. Raising up the suspension in order to install larger tires and gain ground clearance was among the modifications made. The rest, as they say, is history.

Love them or hate them, there's no denying that the Jeep is a big reason 4x4s are so readily available today, and just like those early days, 4x4 owners are constantly tinkering with them. While Jeeps are certainly going to be discussed in this book, they are not the only popular 4x4 these days. Chevy, Ford, Dodge, Toyota, and other manufacturers have all been building popular 4x4s for over 30 years. Without a doubt the most common major modification made to these vehicles is a suspension lift. No other modification enhances a 4x4's attributes as much as a lift system, and an entire multi-million-dollar industry has been developed to satisfy a 4x4 owner's quest for altitude. However, the suspension designs under 4x4s produced since the '40s range from relatively simple systems with solid axles and leaf springs to incredibly complex Independent Front Suspensions (IFS). To further complicate matters, at least 20 companies currently offer suspension lift systems of various designs and lift heights. To the uninitiated, the options and complexities of all these different suspension systems can be bewildering. This book should help you sort through the morass by taking an in-depth look at the most common suspension designs, and then walking you through modifying them safely and properly.

This chapter discusses the basic whys and hows behind purchasing a lift system. This is for the person that doesn't particularly care about the "nuts and bolts" of a lift, but instead wants the basic knowledge and tools to make an educated decision. Chapter 2 contains a closer look at the individual components that make up a suspension system, as well as some of the technology that goes into developing and producing these components. The subsequent chapters contain an in-depth look at the five most common suspension designs utilized by the original equipment manufacturers and how they are modified for increased height and performance. Lastly, we spend some time looking at a number of high-performance suspension designs out there and provide a primer on some of the common hand-fabricated systems developed for a specific purpose.

Obviously, this book focuses primarily on off-the-shelf suspension systems; if you're interested in fabricating your own, there are other books that are perhaps better suited to your needs. Should you want to fabricate your own suspension, some words of caution: An automotive suspension system is an incredibly complex network of individual components that must all work in concert. Designing and fabricating a safe and effective suspension requires an enormous knowledge base and an intimate understanding of key elements. The amount of work required is much greater than installing an off-the-shelf system, and requires a substantial financial investment that is easily double or more what a purchased lift system costs. Plus, engineering a system that is both stable and performs well is very difficult. In other words, there are only a few ways to get it right, but an unlimited number of ways to get it wrong.

One of the best ways to enhance the off-road capability of a four-wheel-drive vehicle is to increase its ground clearance by installing a lift system. A lift raises the body and undercarriage of a 4x4 to make more room underneath. This allows the vehicle to climb over obstacles without getting stuck. In this book, we take an in-depth look at the more common four-wheel-drive suspension systems, and then show you how they are modified for enhanced off-road utility. But before we dive in to the suspension world, a couple terms need to be made clear:

Defining a Four-Wheel-Drive Vehicle

Up until the 1980s it was clear which vehicles were four-wheel drive and which ones were not. These days there are a lot of cars and trucks being produced that are marketed as four-wheel drive, but are really more all-wheel drive. What's the difference? For the purposes of this book, a four-wheel-drive vehicle (or 4x4 as is commonly used throughout this book) is defined as a vehicle capable of transferring power to all four wheels via a transfer case equipped with a Low Range and a selection that provides a true 50/50 torque split to the front and rear axles (usually marked 4-Hi). An all-wheel-drive vehicle typically does not have a Low Range and usually has a center differential that varies the amount of torque split to the front and rear axles. This distinction is being made because Low Range is specifically for off-road use and is indicative of a vehicle capable of negotiating rough terrain in a controlled fashion. An all-wheel-drive vehicle has the enhanced traction capability of a 4x4, but is more at home in slippery environments such

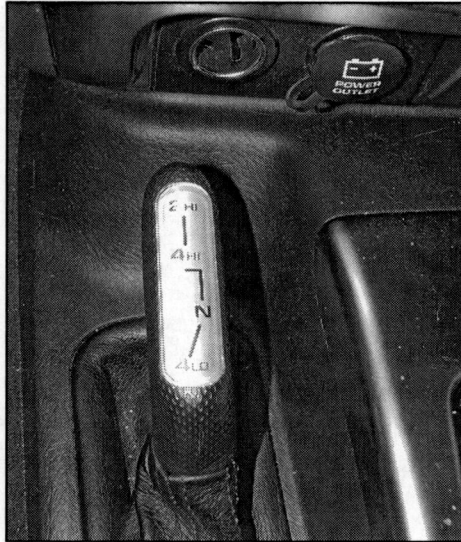

Just because there's not a lever on the floor does not mean it's not a real 4x4! Many late-model trucks have push-button selectors for Hi and Low Range, and some even have a full-time four-wheel-drive mode as well as a part-time selection.

A "true" 4x4 has a Low Range selection, indicated by the 4-Lo on the transfer case lever. Low Range slows the vehicle down and multiplies torque to the axles, allowing better control and increased power during off-road maneuvers. Vehicles without a Low Range are generally more comfortable on slippery surfaces such as ice rather than off-road, and most are based on car platforms. This, combined with the fact that not many AWD vehicles are popular among real off-road enthusiasts, is why not many aftermarket products are available for AWD.

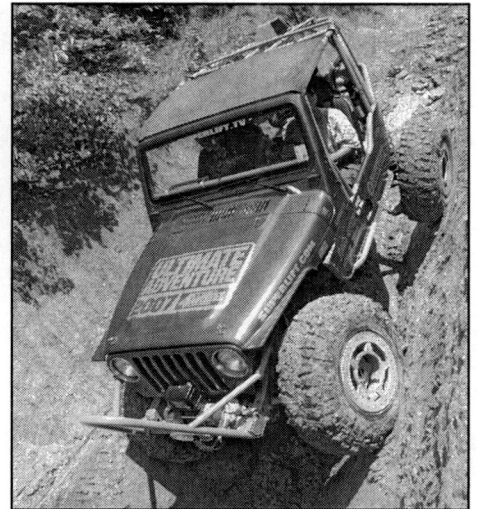

Off-roading is what 4x4s are all about, but if we don't respect the rules on public and private lands, we soon won't have anywhere left to go four-wheeling. As you can see here, this rig is on an established, marked trail. Off-roading does not mean blazing your own trail or cutting your own path in the woods; this is destructive and gives those groups who would rather not allow motorized access to public and private lands more ammunition to deny the rights everyone should have, no matter what their conveyance.

as an icy or rainy road. These vehicles are not well suited for true off-road use and are rarely if ever modified for this purpose. Obviously, we deal with "true" four-wheel drives only.

Off Road vs. Off Highway

Though the term "off road" is clear to many, it is important to explain how this term is used because it is something of a misnomer. Off road is defined as being on any established route that is not paved. Off road does *not* mean blazing your own trail across the desert or through the woods. Though the term "off highway" may be more appropriate, off road has become common vernacular, and as such is the term used in this book. It is every off-road enthusiast's responsibility to stay on established trails on both public and private property. Obey

any posted signs and report any violators to authorities. If we don't follow the rules, we won't have anywhere left to enjoy our sport.

Suspension Basics

By its most basic definition, the suspension system of a vehicle absorbs surface irregularities and prevents them from being transmitted to the passenger compartment of the vehicle. That sounds simple enough, but the fact of the matter is that suspension plays a part in much more than just smoothing out the bumps and dips we encounter on and off the road. Suspension plays a key role in overall vehicle performance; it's what allows a road racer to take a corner at speed without losing control, what enables a drag racer to transfer power to the ground for a

hard launch off the line, and what transforms a ride in a regular car and truck from a bone-jarring experience to an ordinary part of life. But for a four-wheel-drive vehicle, the suspension system is what makes or breaks its safety and performance on and off the highway.

When you mention suspension to most people, they associate it strictly with how comfortably a vehicle rides. That's true, but it also plays a key role in getting engine power transferred to the ground by keeping the tires in contact with the road surface. It locates the front and rear axles of the vehicle, it controls vehicle stability by keeping the vehicle's body and frame steady in corners, and it controls the distance maintained between the vehicle's frame and the ground. So, there is a whole lot more going on here than meets the eye! The fact of the matter is that a 4x4's suspension system must be able to do many things at once. It must be versatile enough for you to comfortably do 70 mph on the interstate, and then lock in the hubs and crawl over rough terrain off road.

The exact components that make up a vehicle's suspension vary according to the suspension's design, but at the most basic there are springs, shock absorbers, sway bars, link or control arms, and compression and extension travel stops. In addition to the suspension components themselves, a variety of other related components must also be included when you're talking about lift systems. These include steering components, the axle assemblies, brake lines, tires and wheels, drivelines, and so on. Each of these topics is covered more in-depth in subsequent chapters.

If you ask any owner of a 4x4 how he or she wants to modify his or her vehicle, you're likely to get "lift kit" as an answer. Why is that? A lift system raises the body and frame of the vehicle to create additional ground clearance. It also provides room for larger tires, which further enhances ground clearance and aids traction. In many cases a lift system also increases suspension travel, providing additional suspension articulation to keep the tires in contact with the ground while climbing over obstacles. Lastly, the motivation for many lift kit purchases is to enhance the look of the vehicle and make it stand out in a crowd. Whether you want enhanced performance or just want that lifted look, it's important to know what's going on underneath before you go shopping for a lift system.

What Lift Systems Affect

In a word, a lift system affects all of the components mentioned earlier. The important thing to remember here is that the axle assemblies are being moved farther away from the chassis of the vehicle, so everything that attaches to both the axles and the frame must be adjusted or corrected. The degree to which each component needs to be modified depends largely on the amount of lift. With a 2-inch lift, for example, very little may be required, while an 8-inch lift on the same vehicle requires addressing each of the items outlined below. Individual components are identified and discussed in Chapter 2, but here is an overview of the basic systems and components affected by a lift system.

Steering linkage / geometry

Perhaps the component most profoundly affected by a lift system, the steering linkage and its operating angles are critical. A lift *increases* steering linkage operating angles, and if left unattended these increased angles lead to a multitude of handling problems. These include bumpsteer, wandering, and in severe cases, linkage bind and premature failure. A variety of different methods correct the steering geometry depending on the linkage design, from dropped pitman arms to replacement centerlinks. The goal with any correction method is to bring the steering geometry as close to factory specifications as possible.

Even at a quick glance at this Twin-Traction Beam (TTB) front end, you can see several parts that have been altered to gain lift. The axle beams have been lowered via bracketry, there are new coil springs and shocks, the brake lines are longer than stock, the steering has been significantly altered, and even the sway-bar has been lowered. A properly designed lift takes into account all important vehicle functions that are affected by altering the vehicle's ride height.

The steering linkage is what transmits driver-controlled input from the steering box to each knuckle. The linkage configuration varies from vehicle to vehicle, but it is critical that their operating angles remain as close to factory as possible or several bad steering and handling problems could result. The method for correcting the steering to work properly with a lift system varies, but this particular Jeep's linkage was corrected with a dropped pitman arm.

Addressing driveline angles is just as important as anything else when it comes to lifting a vehicle. In the case of mild lift systems, little if anything is necessary as drivelines can be fairly forgiving. Even at 6 inches of lift, the rear driveshaft of this vehicle is stock. However, degree shims have been installed beneath the rear springs to tilt the rear axle pinion upward in order to relieve the operating angles with the lift.

Driveline angles

The drivelines, or driveshafts, of a 4x4 transfer engine power from the transfer case to the differentials. Every 4x4 has one for the front axle and one for the rear. When thinking about drivelines, you should be concerned with both proper length and operating angles. In the majority of cases, driveline correction is incorporated into the lift system design via degree shims (in the case of leaf springs), tapered lift blocks, or changing the rotation of the differential via drop brackets on an IFS system. Driveline length is rarely a factor at 4 inches of lift and lower, while some applications can handle 8 inches of lift without driveline modifications. As a general rule, the shorter the wheelbase, the greater the likelihood for driveline modifications. Front driveshafts are more problematic than rear shafts; they are

Link arms (or radius arms as this example shows) are what properly locate the front axles of many vehicles. Link angle is often adjusted by lowering the arms attachment point at the frame via a drop bracket, as this example shows. With other applications, replacements links are needed.

usually shorter and their operating angles increase more quickly than longer rear shafts as lift height is increased.

Link arm geometry

"Link arm" is a broad label for anything that locates the axles under the vehicle, whether it's a radius arm,

---

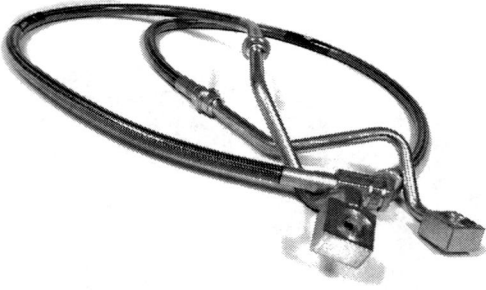

Replacement brake hoses are commonly made of braided stainless steel and are anywhere from 3 to 12 inches longer than stock. The better ones have "leaders," or formed metal sections that match an OE line and help guide the flexible hose away from the tire and other hazards. Also be aware that not all aftermarket replacement hoses are DOT-approved.

control arm, track bar, or panhard bar. The operating angle of these link arms increase as the vehicle is lifted, which affects handling, driveability, and in many cases alignment. Link-arm correction takes many forms, including raising or lowering attachment points, replacement links, or lengthening the distance between the link arm's attachment points.

Brake hoses / ABS wiring

Just as critical as steering geometry, failing to address proper brake hose length is inviting disaster. Ironically, this is one area where people try to cut corners. Brake hose correction takes one of two forms, either lowering the brake hose attachment point at the frame or replacing the factory hoses with extended length pieces. In the case of replacement hoses, buyers beware, as not all replacement hoses are DOT-approved. On late-model vehicles, extending the ABS (anti-lock brake system) wiring is also required. Most lift systems accomplish this by re-routing the wiring or obtaining additional slack elsewhere in the system.

CV axles

On vehicles equipped with Independent Front Suspension (IFS), the CV axles transfer engine power from the differential to the wheel hub. (The differential is rigidly mounted to the vehicle.) Much like drivelines, excessive CV axle angles cause premature wear and failure. CV axle angles are corrected by lowering the differential housing and control arms.

Alignment

The mark of a quality lift system is the ability to achieve and hold factory alignment specifications. Alignment is addressed by a variety of methods depending on the suspension design and is influenced by a multitude of factors. Alignment is discussed at length in the subsequent chapters.

Ride quality

The most arbitrary of the factors listed here, ride quality is influenced by both the obvious factors, such as spring rate and shock valving, and more subtle items such as sway bar pre-load, alignment, and the tire-and-wheel package. Most lift system manufacturers try to mimic factory ride quality in their designs while correcting factory deficiencies such as excessive body roll, brake dive, and vague or spongy driving characteristics.

Common Misconceptions Regarding Lift Systems

Lift kit design has come a long way from the early days of front lift blocks, re-arched springs, and backyard engineering. Lift companies today use the latest in computer modeling and modern manufacturing techniques in order to develop the products they sell. Some of these

On vehicles with IFS, CV axles transfer engine power from the rigidly mounted differential out to the tires. In most applications the CV axles turn all the time, even when the vehicle is not in four-wheel drive. For this reason their normal operating angle must be correct or they will wear out and fail. Excessive operating angles lead to torn CV axle boots, vibration, and premature wear. Since CV axles are not particularly strong in factory form, they are a frequent problem if a lift has not been designed properly.

advances are due to the increased complexity of today's vehicles, but much of it has to do with a greater knowledge base and the desire to build the best-performing, safest suspension systems possible. Even with the substantial growth and increase in technology, there are still some very common misunderstandings regarding lift systems and lifted vehicles in general. Let's take a look at the most common:

A lift system stiffens the ride.

While this statement may have been true 20 years ago, today nothing could be further from the truth. The goal for the vast majority of the suspension companies is to maintain factory ride quality at an increased

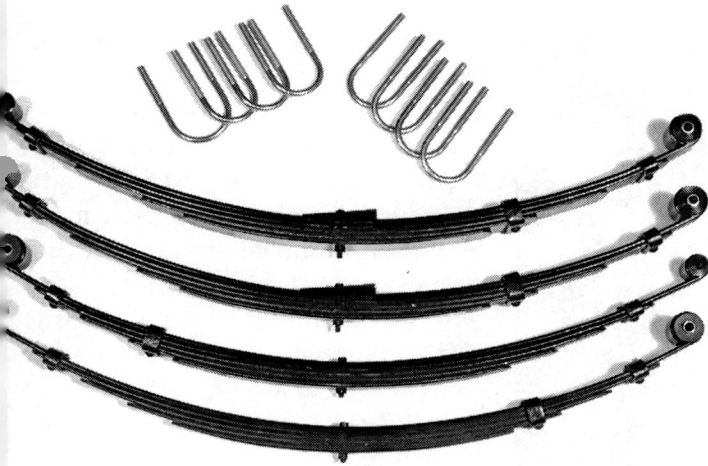

In the '70s, most lift systems consisted of what you see here and little else—four leaf springs. Furthermore, ride quality was never a real factor, so only ride height was taken into account. The results were springs that provided what consumers wanted (more room for bigger tires), but the side effects were pretty brutal in the ride quality department. (Courtesy Superlift)

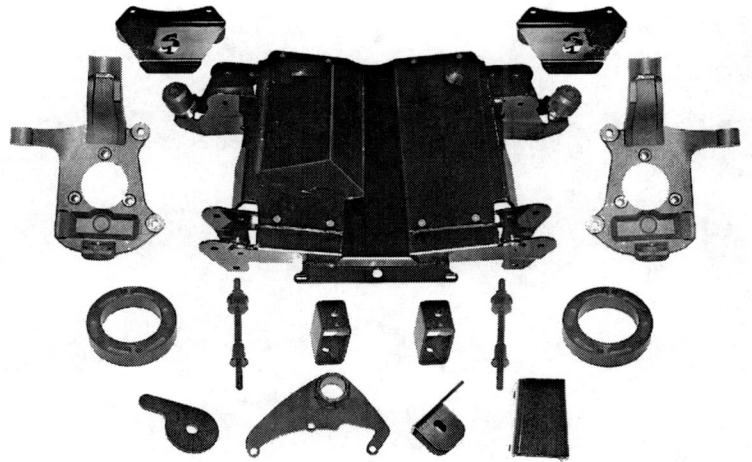

Today's lift kits look very different. The advent of independent front suspension has led to a dramatic increase in a lift kit's complexity, as you can see when you compare it to the previous photo. Furthermore, the market is more competitive these days and consumers are more educated on what a lift kit should have, so ride quality is just as great a concern as maintaining ride height. (Courtesy Superlift)

ride height. They use factory spring rates as a goal, not a starting point. The springs are often sourced from the same companies that build them for the original equipment manufacturers, as are the shocks (springs and shocks are rarely manufactured in-house but instead sourced from an outside supplier). Manufacturers spend hundreds of hours evaluating and tweaking their designs to get the best possible combination of excellent on-road characteristics and good off-road handling properties. Believe it or not, in some cases a lift system actually *improves* ride quality by reducing or eliminating poor factory characteristics.

Lifted vehicles wander all over the place and are difficult to control.

Again, back in the days when a lift kit was just springs this may have been true, but nowadays a quality lift system includes any necessary components to address the steering and alignment. Another factor here is that tire technology has come a long way; once upon a time the only big tires on the market were bias ply, and installing them on a vehicle designed around radials caused strange handling traits. These days there are more radials available than anything else, and tire companies have made great strides in building stable tires in 35-inch diameters and above.

Lifted vehicles are more likely to roll over.

Physics dictate that raising a vehicle also raises its center of gravity. The higher the center of gravity, the more likely a vehicle will roll. However, a lift comes with larger and wider tires, which increases the track width of the vehicle and regains much of that lost stability. So, "go wide as you go tall" is good advice. Still, just like a vehicle that has been modified with a performance engine, a lifted vehicle requires a responsible driver who understands the limitations of the vehicle. Obeying the rules of the road and practicing defensive driving are just as important in a lifted vehicle as they are in any car or truck.

A lifted vehicle can't be aligned and/or wears out tires.

A properly designed lift *will* align to factory specifications. If it does not, the cause is often a problem unrelated to the lift system or an existing problem that the lift and larger tire/wheel package has exaggerated. Worn factory components, such as steering linkage and ball joints, are common culprits. If the vehicle is not in tip-top mechanical condition, a lift will only aggravate an existing problem. Also, modifications to a vehicle rarely stop with just a lift system. Often, heavy-duty bumpers, winches, toolboxes, and snowplows are also part of the mix. All these things add a substantial

It happens, especially when you're pushing the limits off road. Fortunately this was a slow-speed turnover and no one was hurt. The fact is that lifted vehicles handle differently than stockers and require more judicious driving habits. Following the "go wide as you go tall" rule is a good way to avoid ending up on your lid.

Irregular tire wear is a tell-tale sign that alignment is out of whack, which is common to all vehicles and not just those equipped with a lift system. This tire's inside edge is worn more than the outside, indicating that the camber is not correct. Just like all vehicles, a lifted 4x4 must be re-aligned periodically.

amount of weight to the vehicle, which compresses the springs further and lowers ride height. With some suspension designs, changes in ride height have a profound affect on alignment, and if the proper ride height can't be achieved, the vehicle will not align. Doing your homework and/or limiting the weight of accessories that you add to your vehicle negates alignment issues.

A lift kit voids my warranty.

The truth is that no modification you make to your vehicle automatically voids the warranty. The law states it is up to the vehicle manufacturer to prove that the aftermarket component in question caused the failure of a factory component. That's right, the burden of proof is on *them*, not *you*. In reality, some dealerships understand the law better than others, and while one dealership might sell fully lifted and

accessorized vehicles right off the lot, another will attempt ridiculous warranty denials (like voiding the radio warranty because the truck has aftermarket shocks). The best plan is to do your homework and then make an educated choice. If you're having trouble with a warranty claim on your modified vehicle, organizations like the Specialty Equipment Market Association (SEMA) may be able to help. Check out www.sema.org for more information on this subject.

Determining Which Lift System is Right for You

Choosing the right lift system for your needs is not a matter of opening a catalog and finding the cheapest one for your truck. On the contrary, you should consider many factors before you make your purchase, and taking them all into account will undoubtedly lead to a more satisfactory finished product.

There is an enormous selection of lift kits out there for most popular vehicles, but none of them are one-size-fits-all. In an ideal world there would be a lift system that allowed you to take a corner at 50 mph and, at the flip of a switch, unleash gobs of suspension travel for crawling over a hardcore trail. The reality is that suspension has a lot of give-and-take because enhancing the performance of a suspension system in one area usually comes at the sacrifice of performance in another area. Some suspension companies try to achieve a good balance of on-road driving characteristics and off-road performance, while many others design their product around a niche of the market. Answering all of the questions below honestly is a good primer for choosing the right lift system.

What do you want your vehicle to do?

Perhaps the most important question of them all; you need to be sure you answer it honestly. In an ideal world we would have a different rig for every task: one for daily driving, one for towing, one for trial-riding, one for high-speed desert running, and one for pure rock crawling. The truth is that few people have the budget to maintain such a fleet of vehicles and keep the kids in braces and gymnastics lessons. For many, a single vehicle must serve a variety of purposes: the daily driver/family truckster as well as tow rig and weekend 'wheeler. Most 4x4s sold today spend a tiny fraction of their time off road; many are never even put in four-wheel drive. Still, as the multibillion-dollar off-road industry indicates, that doesn't stop people from modifying their vehicles for that "off-road look." There is no shame in modifying a pavement pounder, but the same rules still apply: you need to figure out what is important to you, and then select a lift system that meets those wants and needs.

For a vehicle that primarily sees street use, ride quality should be an important factor in your decision-making. A company that offers engine-specific springs (gas versus diesel) is a good indicator that they have spent some time tweaking the spring rates for good ride characteristics. Progressive-rate springs (discussed in Chapter 2) and application-specific shock valving are also good indicators. While you may want a big lift, erring on the conservative side of lift height may make more sense, as climbing in and out of a tall truck day after day can get tedious. Also, street use can negate the need for costly accessories for your vehicle. Using lift blocks in

The biggest factor in determining which lift kit is right for you depends on how your truck is used. A clean street truck like this one does not need parts to enhance suspension travel, as that really only comes into play during off-road situations. However, optional accessories that enhance ride quality and stability, such as dual steering stabilizers and proper relocation of sway bars, are important.

the rear is a good way to maintain factory ride quality and save money, and multi-shocks are really unnecessary for street use except for looks enhancement.

If you're lucky enough to have a dedicated off-road vehicle, other factors come into play. Looking specifically for lift systems that enhance suspension travel is a good general rule. Many off-road dedicated lift systems have replacement link arms with high articulation joints built in to free up suspension movement. A taller lift system also enables the use of larger tires and greater ground

clearance, both of which aid in off-road utility. Some systems even go so far as to relocate or raise suspension attachment points at the frame. This reduces the number of things that can get snagged by obstacles. If high-speed fire-roading is your thing, then accessories such as multi-shock kits, remote-reservoir shocks, and dual-steering stabilizers are a good choice. In any true off-road rig, replacement springs should be the order of the day. These add expense to the lift system when combined with high-articulation arms, but they also lead to better off-road performance.

For a dedicated off-roader, enhancing the suspension's articulation, or freedom of movement from full compression to full extension, is paramount. Radical modifications, like adding coil-over shocks to the front to replace the factory springs, are all about increasing off-road performance. Things like sway bars don't really come into play, and while a compliant suspension may have more body roll on the street, off the road this same vehicle can be a trail hero. But even though the majority of this vehicle's time is spent off the pavement, it is stable and legal enough to drive safely on the street.

What is your target tire?

Some people focus on a specific lift height and use it to determine what size tire is appropriate, while others fixate on a specific tire size and then add the amount of lift necessary to run them. While neither approach is wrong, the general rule here is to be realistic in your expectations and err on the conservative size. Most lift manufacturers publish a maximum recommended tire size, and some go so far as to recommend a specific wheel backspacing. If you avoid exceeding the lift manufacturer's maximum recommended size, you will have fewer problems in the long run. If you have to run 38s on your Super Duty, be prepared to add 8 inches of lift to your rig in order to clear them; if that amount of lift seems like it's too much, then back down on your tire size.

What is your budget?

Not to be overlooked, budget is another important consideration when choosing a lift system. As a general rule: the taller the lift height, the more expensive the lift system. With a few notable exceptions, there is no such thing as "upgrading" your lift system to a taller lift height. Often, going from 4 inches of lift to a 6-inch lift requires throwing away all but a few components. Therefore, avoid the urge to "settle" for a shorter lift with the hopes of upgrading to what you want later. Either content yourself with what your budget can afford or keep saving until you can get what you want.

While lift height may not be able to be upgraded without starting over, accessorizing your lift system can. Things like dual-steering stabilizers, multi-shock systems, traction bars, skid plates, and dress-up items can all be purchased and installed as your budget dictates. In other words, you can install that 6-inch lift, then go back in six months and add that multi-shock kit and skid plate you've been eyeing. Furthermore, companies that offer lift systems at different levels can be upgraded in stages. For example, Skyjacker offers its Value Flex, Double Flex, and Platinum systems for Jeep Wrangler TJs. The primary differences between the first two are the suspension link arms, meaning you can upgrade the Value Flex system with Double Flex arms. The system can be further upgraded to the Platinum system, which replaces the front coil springs with coil-over shocks. Superlift, Fabtech, and many other suspension companies offer different levels of the same lift height, so you can upgrade to the next level as your budget allows.

Another important consideration here is budgeting *everything* involved with your project. This includes tires and wheels, alignment once the lift system is installed, and labor for the installation if you're not doing the wrenching yourself. If you own an older or high-mileage vehicle, don't overlook the good possibility that you'll need to replace worn factory components, such as steering parts, ball joints, wheel bearings, and so on. If you don't plan ahead for these types of things, you will undoubtedly exceed your budget and anger your significant other if not break your personal bank.

Can the vehicle be returned to stock?

Though this is not an issue for many people, certain mindsets and situations may require the option of returning the vehicle to stock if needed. Some lift kit designs lend themselves easily to being returned to stock, while others are impossible. In applications such as late-model Chevys where a certain amount of frame cutting is needed, some lift manufacturers offer optional return-to-stock bracketry while others don't. If returning the vehicle to stock is even a remote possibility, make sure you hold on to all of the factory components that were removed, and you should be certain that the lift manufacturer has the necessary components to finish the job.

What to Look for in a Lift System

This is a broad subject that is discussed in-depth in the subsequent chapters, but a few thoughts bear mentioning here. First and foremost, you should be looking for a lift system that offers you the "total package," or all of the components needed to lift the vehicle properly. This means everything: from steering to drivelines, from brake lines to rear lift. The best plan of attack is to create a comprehensive comparison chart with specific items that the lifts for the vehicle in question may or may not

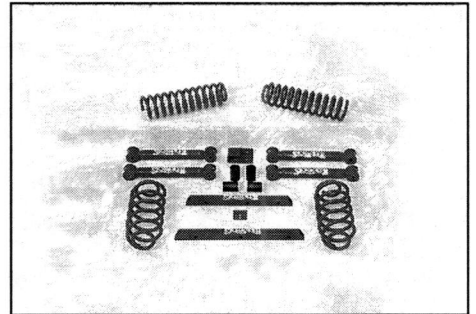

Skyjacker's 6-inch Value Flex system for late-model Jeep Wrangler TJs includes everything needed to lift the vehicle safely. Because things like coil-over shocks and heim joints are expensive, they are substituted for less expensive coil springs and fixed link arms with polyurethane bushings. This system allows budget-conscious consumers to get the desired amount of lift and upgrade as budget allows.

Skyjacker's Double Flex Series 6-inch system includes all the stuff needed to turn your Jeep into a trail hero. While it provides the same amount of lift as its Value Flex kit, there are numerous upgrades that substantially increase suspension articulation. Many of the Double Flex components can be added to the base system as budget allows, making it a good choice for those on a budget or those that want to increase their vehicle's capability as their experience grows.

Some people resign themselves to the fact that they don't care about stability or legality for on-road driving. Buggies such as this one are largely one-off creations where the suspension and body have been radically altered for dedicated off-road use. Though they may be cool, it is difficult to build one of these rigs and have it function well without a lot of knowledge, skill, and experience. (Courtesy Poison Spyder Customs)

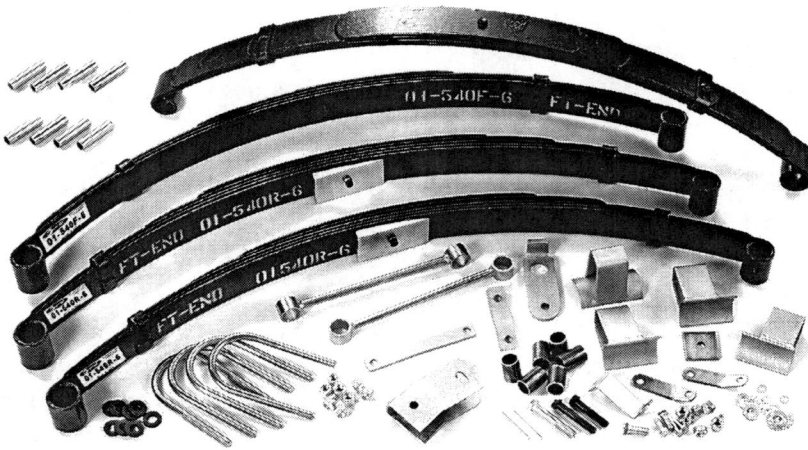

This 4-inch suspension kit for a Jeep Wrangler is commonly listed in mail order catalogs without the pitman arm to correct the steering, which many people agree is required for that vehicle and suspension height. Even though it is cheaper than the equivalent system with a pitman arm, it's not necessarily the right way to go and could end up costing the consumer more in the long run if they have to purchase an arm separately rather than as part of a package deal. Note that shocks, which are also required, are also not included. (Courtesy Superlift)

include. This is the best way to ensure you're comparing apples to apples when shopping.

A typical word of caution here is buyer beware. In an effort to be a price leader, lift manufacturers and retailers often list "optional" components that are actually required. For example, with a 4-inch lift system for a Jeep YJ, some manufacturers include a dropped pitman arm and compression stop extensions automatically, while others don't. Another common trick is to not include shocks in the base kit price, even though they're a must. Fewer components mean less cost, but not necessarily a complete lift system.

A more recent development with the emergence of the Internet is "no-name" or generic lift systems. Often these lifts are cheaper, sometimes hundreds of dollars cheaper. While some may have quality components, ask yourself if they include everything needed and whether or not you'll be able to contact them if you have a problem before you buy. Remember, those brands that are synonymous with the off-road industry have gotten that way through years of developing and selling quality product.

The Domino Effect

Though not directly related to lift systems, there is a phenomenon that often occurs when you start modifying a vehicle called the domino effect. For example, you add a lift and larger tires to a vehicle and notice a performance loss because the factory gears are too tall to get the larger tires rolling as quickly as they did before the modifications. The answer is lower (numerically higher) gears to restore the lost performance. But since the gears are being changed, wouldn't it also be a good idea to add some traction devices (limited-slips or lockers) at the same time? Now that the vehicle has the clearance and gearing, you begin taking it out on the trail. After an excursion or two, it is clear that if you want the sheet metal to stay straight, some heavy-duty bumpers are in order to protect the body. Since you're adding bumpers, why not a winch?

Perhaps you get the picture: once you open the door to vehicle modifications, it's hard to know when to stop. The moral of the story is that individual modifications usually affect other vehicle systems and traits, which in turn lead to more modifications. Some people may look at this as a Pandora's box, but don't be afraid of the domino effect. It is part of what drives us to individualize our vehicles and tailor them to suit our individual needs and uses.

While the lift on this Bronco is readily apparent, what is not is that it has been re-geared with 4.56:1 gears, has lockers, and the entire interior carpet has been replaced with bedliner material, among several other modifications. When you begin the process of modifying a vehicle it is often difficult to know when to stop, and while your checkbook may take a beating, there's nothing quite like building and driving a one-of-a-kind vehicle.

SUSPENSION THEORY

What Goes on Behind the Scenes

Now that we have the basics out of the way, let's take a look at the individual components that make up a lift system and some of the technology that goes into building each piece. Even though dozens of 4x4s are currently on the market, they share many common components. Furthermore, the suspension systems fall into five major categories, each of which are discussed at length in subsequent chapters. These include:

- Solid axle with leaf springs
- Solid axle with coil springs
- Twin-Traction Beam (TTB)
- Independent Front Suspension (IFS) with torsion bars
- Independent Front Suspension (IFS) with coil struts

Each of these major categories is largely its own animal, with its own unique advantages and quirks that are discussed in-depth later. However, a few general rules and theories are common to all of them, and that is what we are discussing here. Fur-

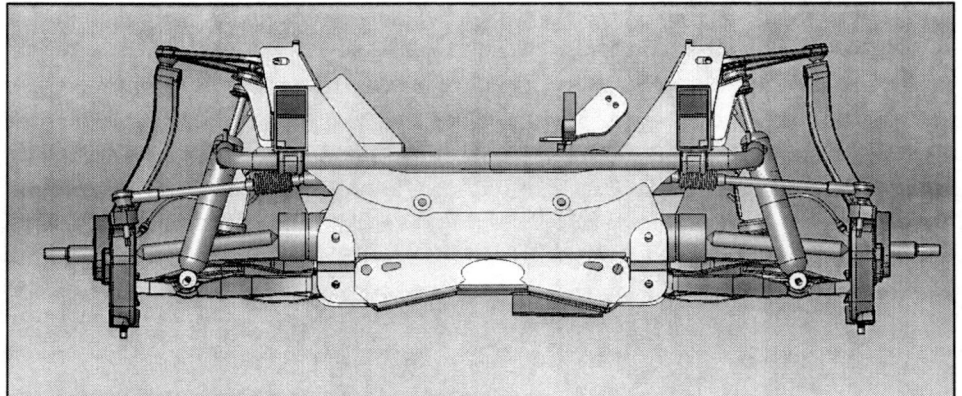

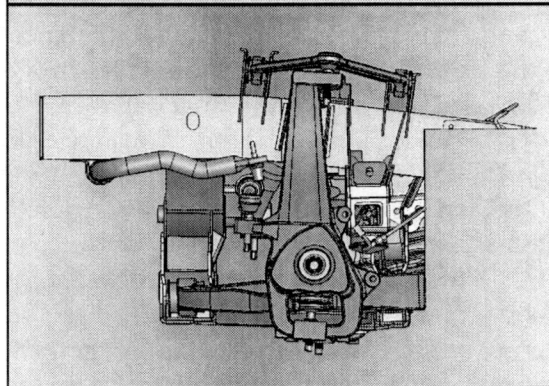

Once upon a time, lift manufacturers used trial and error to design and test lift systems before releasing them to the general public. With complex IFS the norm rather than the exception these days, top lift manufacturers use sophisticated CAD software like Solid Works to design lifts in a virtual world, and employ advanced manufacturing techniques to ensure consistency from one bracket to the next. In a few cases, aftermarket designers use OE files to start designing products before a new model has even hit the car lots. (Courtesy Superfit)

thermore, they use several common components, namely springs and shock absorbers, which are constructed in the same ways. Therefore, we take an in-depth look at how these items are made and what traits among them can potentially impact ride quality and longevity.

With solid-axle systems, the drag link (the steering part running from the steering box to the passenger side knuckle) and the track bar (responsible for locating the axle side to side) must move in the same arc whether the vehicle is stock or lifted. If these two get out of phase with one another due to improper correction with a lift system, all sorts of negative handling characteristics will result. In this case a prototype track bar has been fitted to test the driving characteristics on this prototype 6-inch long-arm system for a Jeep TJ.

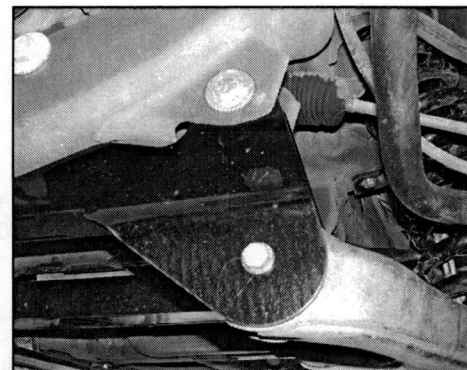

Drop brackets are an important part of an IFS system over 2 inches. Plate steel brackets are utilized to relocate several key components so that they stay within their proper range of motion while working with a suspension lift. In this case, a new front crossmember is used to relocate the lower control arm from its original position on the frame (now serving as the upper attachment point for the crossmember). By moving the control arm and several other components downward, the frame of the vehicle is raised, thus achieving lift.

Suspension Geometry

This is a broad term that can mean many things depending on the suspension system's design. However, it is a term that must be understood when studying suspension systems in general, and geometry plays a critical role in proper lift system design.

In general, the term "suspension geometry" refers to the operating angle (or range) of the suspension components themselves. It refers to the angle of the link arms that attach the axles to the frame in solid axle systems, as well as the angle of the control arms in an IFS system. *The important thing to remember here is that with only a couple of exceptions, a properly designed lift system's goal is to mimic the factory operating-angle of the various suspension components.* Why? The factory engineers designed a multitude of components to work within a certain range of motion, which is the amount of distance or travel from full compression (tire stuffed in the wheel well) to full extension (tire drooped away from the body). If this range is exceeded, many critical components become in danger of exceeding their individual operating range and breaking. These include tie rods, ball joints, CV axles, and numerous other items. In short, it is best to stay within the factory-designed suspension range or very bad things could happen.

How, exactly, is correction of the suspension geometry achieved?

With IFS, the common correction method is via plate steel brackets that lower the attachment point of the factory component in question. Why lower the attachment point rather than raise it? We lower it because a lift system increases the distance between the drive axles and the frame of the vehicle. Since we're *increasing* the distance, we must *lower* the components in order for them to operate properly.

For example, let's take a look at a stock late-model ½-ton Chevy truck. At normal ride height, you can see that the lower control arms are attached to the frame and the steering knuckle (where the wheel bolts into place). Since the vehicle is on the ground with the suspension supporting the vehicle's weight, this is

With solid-axle vehicles the lift method is a little different. The actual lift is achieved by installing taller leaf or coil springs. These springs move the axles farther away from the vehicle, hence raising it. However, installing a lift is never as simple as slapping a pair of springs in place because everything that attaches to both the axle and the frame must also be addressed. This includes link arms, brake hoses, and sway bars just to name a few. Some of the red bracketry that addresses those items is visible.

the normal operating angle for all components, including the steering, alignment, CV axles, ball joints, and so on.

Now let's look at a vehicle equipped with a 6-inch lift system. Instead of attaching directly to the frame, the lower control arms are now bolted to a new crossmember that provides about 6 inches of drop from the original attaching point to the new one. The differential has also been relocated to correct CV axle angle, and a new steering knuckle has been installed to correct tie rod angle and span the increased distance between the upper and lower control arms. Thanks to the

drop of these and other components, all of the critical components operate at close to the same angle as they were originally, which is the ultimate goal for lift manufacturers.

One last look at the same suspension design reveals that even with no weight on the suspension, the suspension components remain within proper operating specs. If these angles were to be seen at normal ride height, at full extension everything would be way beyond its designed range of motion.

It should be noted that most suspension components have an operating *range*, so there is a certain amount of "wiggle room" for correcting operating angles. Using our same example, many lift manufacturers lower front suspension components 5 inches with a 6-inch system. There are many reasons for not going that extra inch, most of which involve discussions beyond the scope of what we're focusing here, but the short answer is that they are working within a *range* of motion, and the smart ones don't exceed factory specifications. They might make up for the difference by moving the extension travel stop (this is what "shuts down" the suspension and prevents it from traveling further) or altering geometry in other areas. Just don't freak out if

A stock IFS Chevy at normal ride height.

A lifted IFS Chevy at normal ride height.

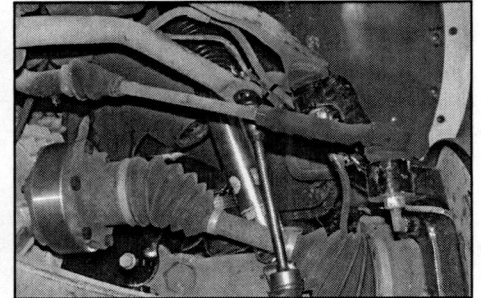

A lifted IFS Chevy at full extension travel.

Controlling extension and compression travel is critical with any IFS design, and well-designed systems pay very close attention to both. In this case, a bracket has been supplied to alter the extension travel stop to compensate for other modifications associated with the lift system. Though it is tempting to change both stops in an effort to improve overall suspension travel, this is best avoided with IFS systems due to their narrow factory operating range, which protects key factory components.

vehicles are very sensitive to changes in alignment, and, to a lesser extent, wheelbase. Therefore, while the safe operating range of the link arms is more than enough to accommodate a mild 4-inch lift, changing ride height adversely affects alignment, hence the need for a replacement link that addresses the needed correction to work with the lift. The moral of the story here is that correction can take many forms, and many components correct more than one issue that results from raising the vehicle's ride height.

It was mentioned earlier, sometimes lift system designers radically alter the suspension geometry, and in all cases these alterations provide significant improvements in suspension travel. In most cases, these alterations involve replacing short factory link arms with much longer arms. The reason is simple enough and has to do, once again, with operating

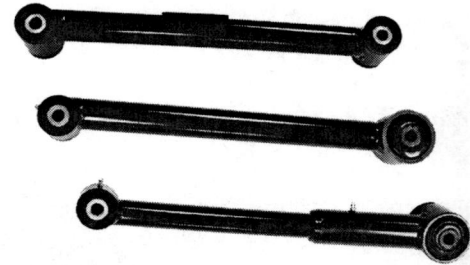

Replacement link arms take many forms, but they all have one thing in common: they correct caster as well as wheelbase change with a lift system. Some are strictly replacements for the factory components (top), while others provide adjustment and absorb the "twist" that develops as the suspension is loaded on one side (bottom). The ability to twist rather than bind within its attachment points can add a significant amount of travel and better off-road performance as a result. (Courtesy Superlift)

you find your 6-inch system drops the components a little less than advertised!

Another method of correction commonly used is to replace a factory link with a piece provided in the lift system. A good example of this correction method is a 4-inch lift system on a Jeep Wrangler (TJ). Virtually all proper lift systems include replacement lower link arms for the front suspension, and some include rear arms as well. With this application, correction via the replacement links is more a function of alignment (in the front) and driveshaft correction (in the rear) than correcting excessive operating angles. Generally speaking, solid-axle vehicles that utilize link arms are very forgiving when it comes to increasing their operating angle, but those same

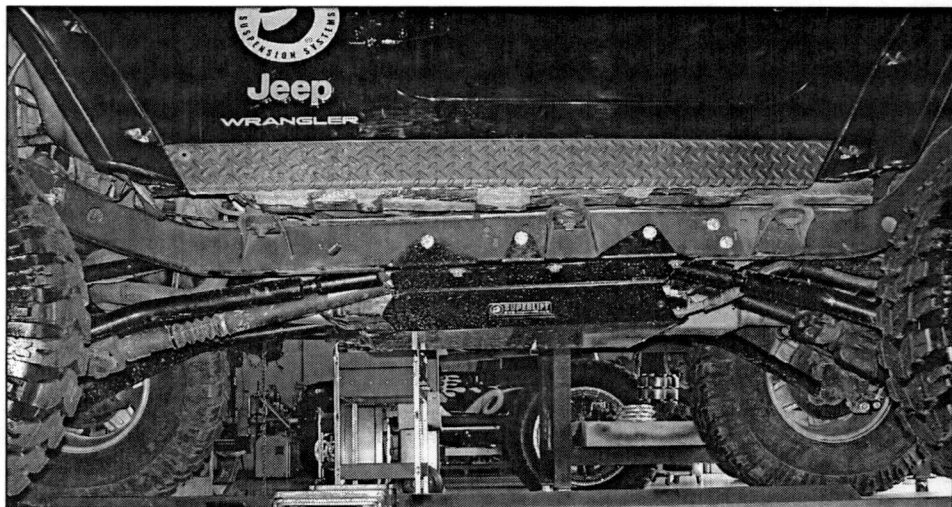

Long-arm systems have become the standard for hardcore off-roaders. They replace the short factory links with substantially longer ones that usually connect to a new transmission crossmember. Operating on the same basic theory as extended radius arms for Fords, providing a longer arm for the suspension means less dramatic changes in geometry as the suspension moves through its travel cycle, enabling a greater amount of overall movement as a result.

angles. The most common example these days is to replace the short factory link arms on a Jeep Wrangler (TJ) with arms that extend all the way to the transmission skid plate. The operating angle of these longer arms does not change as dramatically as the short arms they replace, as the suspension cycles. This means they allow more movement before they begin to bind, and they also don't "roll" the axle as much from full extension to compression, so less caster change is evident. The end result is dramatic improvements in suspension articulation and the ability to handle more lift safely. Though expensive, long-arm systems are hard to beat for hardcore off-road use.

Alignment Explained

Regardless of the suspension design, all vehicles must follow a common set of rules in order to drive and steer properly. Though the exact components that control these rules vary among the different suspension designs, they all share a common goal. These rules fall under three major categories:

Caster

Caster is the fore or aft slope of the steering axis. The steering axis is a line drawn through the upper and lower ball joints of the knuckle. Positive caster is when the bottom of the steering axis line is in front of the tire's contact patch. Zero caster is when the steering axis is at 0 degrees, or even with the tire's contact patch. Factory alignment specs for basically all vehicles call for a certain degree of positive (shown) caster.

Caster has a profound impact on many driving traits. It helps keep the vehicle pointed straight ahead

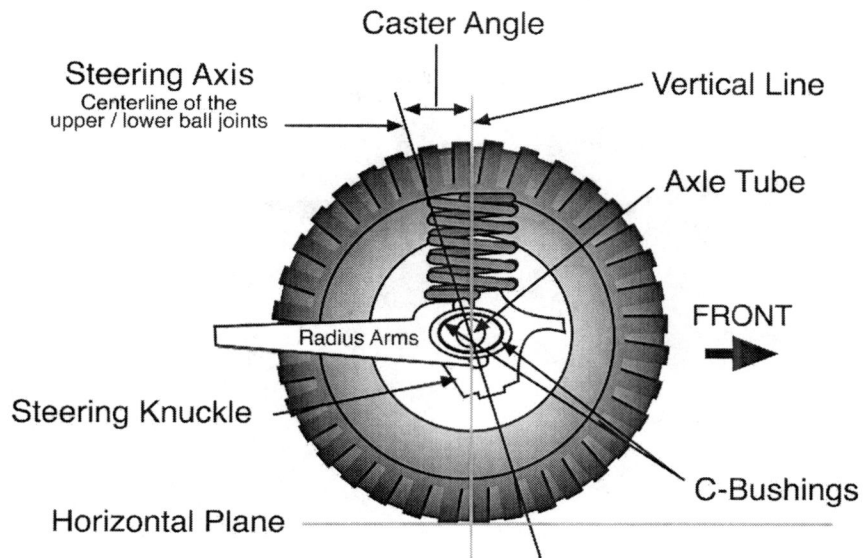

Caster Angle. (Courtesy Skyjacker)

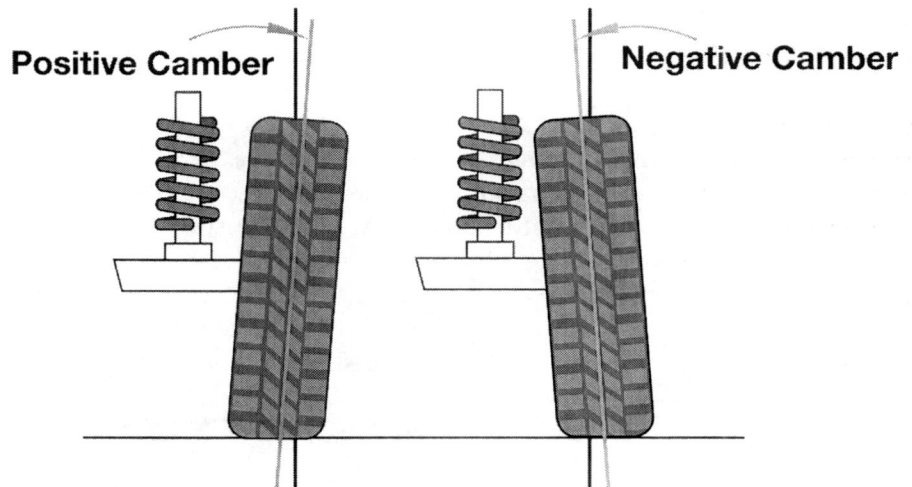

Camber Angle. (Courtesy Skyjacker)

instead of wandering off one way or another all of the time. It also promotes the steering system's ability to self-center. If caster is wrong or negative, just the opposite will occur.

On some leaf-spring/solid axle designs, caster is set by the axle, and very little needs to be done to correct it with a lift system as long as degree shims are not used at the spring pads to improve driveline angle. Other vehicles are very sensitive to caster changes, including most solid-axle

systems that utilize link arms or radius arms to locate the axle.

Camber

Camber is the inward or outward tilt of the front tires as viewed from the front of the vehicle. Inward tilt is negative, outward tilt is positive. Camber is used to distribute load across the entire tire tread and can impact cornering as well as tire wear. If the camber is wrong, the inside or outside edge of the tires can wear

Toe Angle. (Courtesy Skyjacker)

unevenly. If camber is out on one side, it can cause the vehicle to pull to the side with more positive camber.

Camber is generally problematic on TTB Fords and IFS vehicles. With TTB vehicles, camber is corrected via axle pivot brackets and special shims positioned under the upper ball joint. Spring height also plays a key role in camber with TTB. With IFS systems, camber is typically adjusted via cam bolts that attach the upper control arms to the frame. Camber is largely fixed on all solid-axle vehicles.

Toe

Toe is the side-to-side difference in distance between the front and rear of the front tires. If the distance is closer at the front, it's called toe-in. If the difference is closer at the rear, it's called toe-out. By far the simplest of the three alignment specs to correct: typically tie rod length is adjusted. Generally speaking, a vehicle needs about 1/16 inch of toe-in. If the toe is wrong, the telltale signs are uneven tire wear, wandering, and a steering system that hesitates to self-center.

All three of these specifications must be right or a horrible array of driving characteristics will result. Even just one a little outside of proper specs can lead to irregular handling traits and tire wear. The

specs for all three vary from vehicle to vehicle, but the important thing to remember is that unless an aftermarket suspension company provides specific alignment specifications, the vehicle should always be aligned to whatever specs the OE manufacturer indicates. Alignment shops sometimes don't understand this and demand that the customer supply the necessary numbers. If this happens it is clear that the shop in question is either inexperienced or unwilling to align lifted trucks, and you should think twice before paying for their services. Because it's so important to

get the alignment right, it's best to follow the recommendations of the local off-road shop if they don't have the equipment to do it themselves.

One very recent trend has developed related to alignment and the electronic stability control systems on the newest vehicles. These systems utilize sensors that monitor steering wheel position, and if the steering wheel is not perfectly centered when the vehicle is aligned, it will activate a false malfunction code and the associated warning light. If the vehicle is equipped with an electronic stability control system, it is best to have the vehicle aligned using modern laser-equipped machines.

Leaf Spring Tech

Leaf springs are the single most commonly used suspension components on trucks today and are certainly the oldest. A leaf spring is one of the simplest spring types that can be found and is quite possibly the most durable, which explains its popularity to this day. Leaf springs

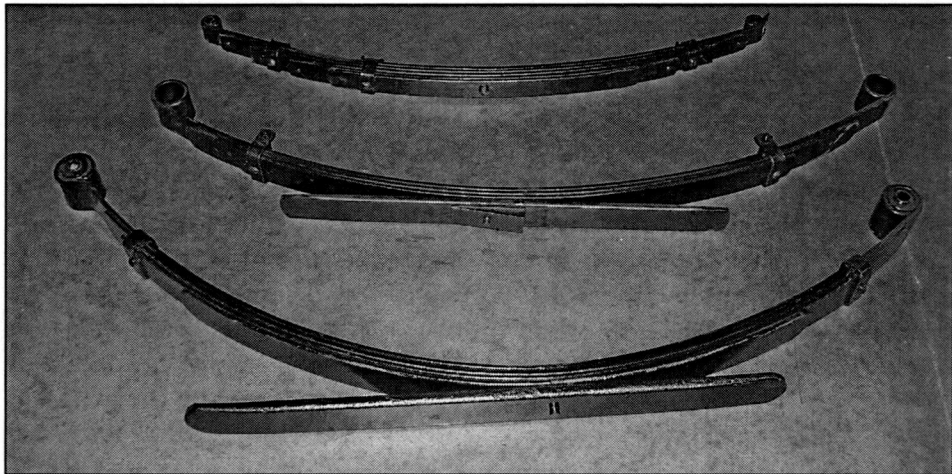

Leaf springs come in a variety of lengths, widths, thicknesses, and arches. With the bewildering array of different leaf spring designs out there it can be difficult to determine which one will best serve the preferences of the owner and the vehicle's use; note the two are not necessarily the same.

are versatile, relatively compact, and can serve double-duty as both a suspension attachment point and a load-carrying member. But even though leaf springs are simple at first glance, there is a surprising amount of technology and know-how that goes into building a spring that provides good ride quality, load-carrying capacity, and longevity.

Basic Anatomy

A leaf spring starts with a single leaf with an eye on either end called the main leaf. Beneath the main leaf is a series of support leafs of varying thicknesses and numbers depending on the amount of weight the vehicle is designed to carry. At the bottom of

The thickness and number of individual leafs that make up a spring pack vary according to the design of the spring and its intended load-carrying capacity. Most truck rear springs have a thick overload leaf at the very bottom of the spring pack that is flat, unlike the others. This overload leaf has no impact on ride quality until the pack compresses enough to rest against it, at which point it provides extra load-carrying capacity. One trick the hardcore off-roaders use is to remove the overload leaf from the pack. Though this compromises its ability to support heavy loads, it does free up suspension travel.

many rear springs is a thick flat leaf called an overload leaf, which provides additional load-carrying capacity once the spring flattens enough to engage the overload spring. Other basic leaf spring designs may have a top-mounted overload leaf in which a curved spring mounted above the main leaf engages two tabs on the frame when the suspension "squats" due to additional weight being placed on the vehicle (such as the tongue weight of a trailer or an in-bed camper).

A leaf spring performs multiple roles: it must locate the axle of the vehicle, absorb surface irregularities instead of transferring them to the passenger compartment, and support the load of the vehicle plus any cargo. Unfortunately, enhancing a spring's attributes in one area usually comes at the cost of its performance in another; generally speaking, the greater the load-carrying capacity of a spring, the stiffer the ride. Conversely, providing a softer spring to optimize ride quality reduces load-carrying capacity and compromises the spring's ability to properly locate the axle by being susceptible to axle wrap (this is discussed in detail in Chapter 8). Designing a leaf spring to perform well in all of its roles is a delicate balance of give and take with all suspension designers, whether OE or aftermarket. In the context of lift systems, designers primarily focus on ride quality and longevity of the spring, or its ability to provide comfort and maintain proper ride height.

Defining Spring Rate

There is one term used when talking about leaf springs that should be understood by anyone shopping for a lift system: spring rate. It is defined as the amount of weight it takes a

A leaf spring's job is a hard and often thankless one in any suspension system. Because it must locate the axle and supply the necessary suspending duties, leaf springs are subjected to more stress than other components. Designing one to perform well in all of the required tasks is a delicate balance for any spring designer.

spring to deflect 1 inch at a given height. Spring rate is essentially the measurement of the spring's load-carrying capacity, but the numbers also provide a glimpse at how well or poorly two or more springs will ride by comparing their respective spring rates. The lower the number, the better it rides, since it takes less weight to move the spring. However, buyer beware—there is, unfortunately, no standardized spring rate measurement utilized among *all* of the lift manufacturers. The most common and accepted method of measuring spring rate is to determine the amount of weight it takes to deflect a spring 1 inch *at normal ride height*, that is, with the curb weight of the vehicle already applied to the spring. But, not everyone follows this rule and the testing methods are far from universal. The only way to accurately compare spring rates is to first verify that the manufacturer's testing methods are identical. Also, keep in mind that lower numbers are not necessarily better if load-carrying capacity and longevity are primary concerns.

Leaf Spring Ride Quality: A Function of Design

In the early days of off-the-shelf lift systems, the only focus of designing a leaf spring was increasing ride height; ride quality was far down on the list of priorities, if it was considered at all. Thankfully those days are gone, and in today's ultra-competitive market the primary focus of lift system designers is optimizing ride and handling characteristics. That being said, the fact of the matter is that there have been no substantial technological advances for leaf springs in decades; there is no magical potion that creates the perfect leaf spring. Therefore, the designers of lift springs must balance their requirements so that the leaf springs provide optimum ride quality while *also* providing a reasonable amount of longevity. That is, a designer can easily create a soft-riding spring that provides unsurpassed comfort only to have it lose arch and sag because there is not enough spring rate to hold up the weight of the vehicle for a long period of time. Conversely, they could design a spring that lasted for the life of the vehicle and never sagged, but its rate would be so high that lifted truck owners would have their chiropractor on speed dial.

Determining the perfect optimum spring rate is largely a proprietary process among the different manufacturers and involves a certain amount of voodoo science as well as trial and error. One of the biggest factors in ride quality is reducing friction within the leaf pack. You see, as a spring compresses and droops, the individual leafs that make up the

spring pack must slide against one another. If they are not allowed to slide easily, ride quality suffers. As a result, a few easily identifiable design elements can be built into a leaf spring to aid ride quality and let you know it is a well-designed spring:

Squared-off or blunt-cut leafs present on a spring pack typically indicate the spring is designed more around carrying a load and maintaining proper ride height than providing optimum ride quality. Blunt-cut leaf springs were the standard in the early days of lift kits, and this is most likely the source of how lift systems developed an early reputation for ruining the ride quality of a vehicle. Today, blunt-cut springs are the exception rather than the rule.

Tapered leafs are an indicator the spring designer has paid close attention to ride quality. The taper of each leaf allows them to slide more easily against one another.

Tapered vs. blunt-cut leafs: Probably the most logical and easily identifiable; take a look at the ends of each individual leaf in a spring pack. If each one has a squared off end, this is an indicator that the spring will not ride well. Springs with tapered ends are shaped that way to reduce the amount of friction between their fellow leafs. Think about it: Is it easier to push something up a sharp curb or a ramp?

Anti-friction pads: Evident on more expensive springs generally of a higher quality, the use of anti-friction pads increases the cost to manufacture a spring, but greatly reduces inter-leaf friction. These plastic-like pads are attached to the ends of each leaf and act as a lubricant between the leafs, which allows them to slide smoothly against one another.

Cinch-style vs. bolt-style clamps: Every leaf spring must have one or more clamps around it in order to hold the pack together; if it doesn't, the leafs fan out and the spring pack

If you look closely you can see the anti-friction pads installed at the tips of each leaf in this spring pack. These pads are another way to reduce inter-leaf friction. Combined with tapered leaf ends, friction pads are indicative of a spring that provides better ride characteristics.

comes apart. However, not all clamps are created equal: A cinch-style clamp tends to apply pressure to the spring pack and increases friction. A bolt-style clamp does its job of holding the pack together without increasing friction. It even has the added bonus of increasing suspension droop in some cases because it allows the support springs and main leaf to separate from one another when unloaded. Though a bolt-style clamp is slightly more costly than a cinch-style clamp in manufacturing, they have the advantage when it comes to ride quality.

Military wraps: Often a sign of a high-quality spring pack, a military wrap is a design element in which the secondary leaf (the one directly under the main leaf) extends out past the main leaf and wraps around the spring eye. This is a safety feature; in the event the main leaf fails, the secondary leaf holds the spring pack together so the vehicle can be limped back to town for proper repairs. The military wrap also provides greater support to the main leaf by supporting it across its entire length. It is common for very high-quality spring packs to have military wraps at one or both ends.

Thick vs. thin leaf packs: Quite possibly the most misunderstood quality of a spring pack, the relative thickness of the individual leafs within a spring pack and the overall thickness of the pack is not necessarily an indicator of a spring's ride quality or load-carrying capacity. The thicker the spring pack, the stiffer the ride, right? Wrong. Some 1-ton trucks use two- or three-leaf packs, while ¼-ton Jeeps use packs with seven or more leafs. In fact, there are many factors involved here. When comparing two different

How the spring pack is held together also has an impact on ride quality. Cinch clamps tend to hold the spring tightly together and increase the amount of friction within the spring pack, while bolt-style clamps do their job without impacting ride quality. One ride-quality trick is to open up cinch clamps slightly to allow the leafs to move more freely against one another. Just don't open them up all the way.

A military wrap is a worthy addition to any quality spring pack because it continues to hold the spring pack together in the event of a main-leaf failure, in which case the vehicle can be limped home rather than stranding the driver. Military wraps also provide additional support to the main leaf. A ¼ military wrap offers the same support without the safety feature. Military wraps are always an indicator of a high-quality spring.

So, was this spring designed around ride quality or load-carrying capacity? The answer is the latter and the clues include multiple thin leafs, bolt-style clamps, tapered leaf ends, and anti-friction pads. Never judge a spring by its thickness alone.

Polyurethane bushings offer better handling traits and are favored in colder climates because they are immune to road salt and other debris. However, urethane bushings also tend to squeak and can negatively impact ride quality. Factory-style rubber bushings are generally better for ride-quality concerns.

Comparing stock and lifted springs is common in magazine articles because the difference is often dramatic. However, the increased arch means a reduction in ride quality and load-carrying capacity. This is one reason why replacement rear springs are not usually the best choice for a truck owner who wants to maintain the factory ride and load-carrying capacity.

springs for the same application, however, generally speaking a spring with relatively few thick leafs will not ride as well as a spring with multiple thin leafs. This is because the pack with multiple thin leafs has a more even and progressive rate as it compresses, while the thicker leafs create spikes in the spring rate curve. The moral here is not to judge a book too much by its cover.

Bushing material: It is not a coincidence that virtually all of the vehicle manufacturers use rubber bushings in their leaf springs, while the majority of the aftermarket springs use polyurethane bushings. Speaking strictly in terms of ride quality, rubber bushings are superior because they allow more give and movement than poly. However, poly bushings can reduce play and sponginess in the suspension, making them a good choice in applications where firming up the ride of the vehicle to improve its handling is desired. Suspension manufacturers also point to poly

bushings being less susceptible to the elements than rubber, while longevity is about the same. However, poly bushings do tend to squeak if not lubricated regularly, and it is not a coincidence that poly bushings are more prevalent in aftermarket springs while being less expensive than rubber to manufacture.

Spring arch: Suspension engineers agree that the best-riding, best-performing leaf spring is a long, flat one. This is because a flat spring has less friction between its individual leafs. Also, a longer spring allows more deflection or movement from its normal shape. This is why rock crawlers often install longer springs on their vehicles. The longer the distance between the attachment points of the spring, the less arch the spring needs to have at a given ride height and the more suspension travel it can provide. Because arch is an important factor in ride quality, one can expect reductions in ride quality as lift height increases.

Coil Spring Tech

Around nearly as long as leaf springs, coil springs represent something of a compromise in "old school" technology in that they eliminate a few of the known problems with leaf springs while increasing a suspension system's complexity. Coil springs have been used for decades by the majority of vehicle manufacturers, in some cases returning to them after experimenting with other suspension types.

The coil spring itself is a simple device: a single length of wire coiled into a desired diameter and height. Because there is space between the individual coils of the spring, there is no friction present; all of the resistance in the spring comes from the tension within the wire making up the spring. However, a coil rarely wants to stay parallel relative to the load being placed on it. If allowed it will move sideways or perpendicular to the load, requiring tight controls

Coil-spring-equipped trucks are the common denominator today, with two of the "Big Three" truck makers using them on their ¾-ton and heavier trucks. Plus, Jeeps have been utilizing them for 10 years now. Coil springs generally offer superior ride quality and often provide more flex than leaf-spring variants.

to keep it where it is supposed to be. Also, unlike a leaf spring, a coil spring cannot be relied upon to serve as a locating member of the suspension, so there must be other links and attachment points present in order to keep the axle located properly under the vehicle.

Because coil springs are so simple there is not a lot of difficulty figuring out what makes them tick. They continue to be a popular choice among both vehicle engineers and custom fabricators because of their simplicity, low cost, and performance traits.

Standard vs. Progressive

Perhaps the most common terms associated with coil springs are "standard" and "progressive" rate springs.

Having an equal distance between the wraps of the coil identifies standard-rate coil springs. This means that the coil compresses evenly as weight is added. By no means a negative, standard coils come straight from the factory on many new vehicles and perform well in just about every situation as long as it has the proper spring rate. The majority of the lift systems available today utilize standard coils.

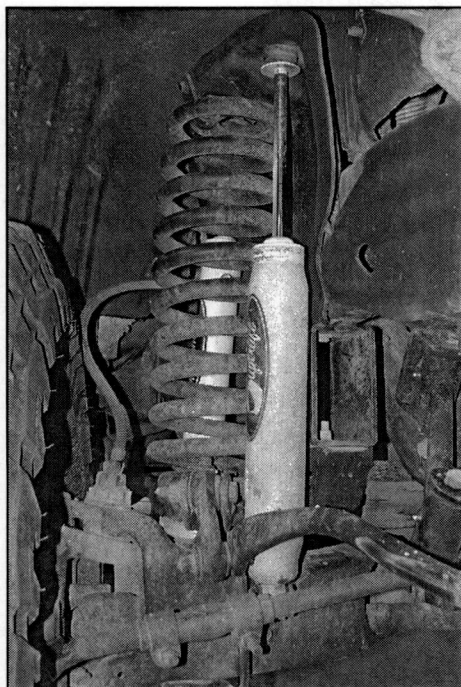

Having one section of coils closer together than the others identifies progressive coil springs. This allows the spring to "ramp up" the rate when compressed to a certain distance while providing a more comfortable light rate under normal street conditions. Progressive coils can also be slightly conical in shape. (Courtesy Superlift)

As the terms imply, a standard coil has an even spring rate curve as it compresses. Standard coils have wraps spaced evenly apart, and as weight compresses the spring, its rate goes up in a linear fashion. Despite marketing campaigns that indicate otherwise, the vast majority of aftermarket springs today are considered standard. A properly designed standard coil provides years of reliable service and good ride characteristics in all environments.

Progressive-rate springs are generally found in more expensive suspension systems and it is easy to identify them. While a standard coil has wraps spaced evenly apart, a progressive coil's wraps gradually get closer together. This allows one section of the spring to do the majority of the work at a certain height. Then, as the coil is compressed more, another section of

---

Many lift manufacturers offer two or more shock options with their lift systems. Most come standard with conventional hydraulic shocks, with gas-pressurized versions available as upgrades. In addition to those two options, Superlift also offers a remote-reservoir upgrade with select lift systems. (Courtesy Superlift)

An inside view of Bilstein's 5100 series shock absorber. A typical monotube shock, Bilstein uses a floating piston to keep the oil and gas separate, which helps reduce shock fade. (Courtesy Bilstein USA)

the coil kicks in and provides additional rate for load-carrying capacity and stability. Coil springs that change diameter across their length (i.e., narrower towards the top or bottom) are also progressive.

Shock Absorbers

Without question, the items that can have the most dramatic impact on ride quality, handling, and overall suspension performance are the shock absorbers. They can make a good suspension system better, but they can also ruin a great one if not valved properly. Choosing the proper shocks is a lot like picking out the perfect set of tires: there are a lot of choices out there and the criteria is usually different from vehicle to vehicle and person to person.

How It Works

The purpose of a shock absorber is pretty well defined by the name; it controls the sudden compression and release of the suspension. When a road irregularity like a pothole is encountered at speed, the suspension system compresses suddenly once it's hit, but this sudden buildup of spring energy wants to release just as forcefully. Shocks help control that release by converting kinetic energy (suspension motion) into thermal energy (heat). But to understand how that happens we need to take a look inside a shock.

A shock consists of a cylinder or reservoir filled with oil and a rod with a piston on the end of it. The piston is submerged in oil and has a tight seal with the walls of the cylinder just like a piston in an engine. In the head of the piston are tiny holes covered by flat metal discs. When pressure is applied to the shock rod, the hydraulic force causes the metal discs to bend away from the holes and allows the oil in the cylinder to pass from one side to the other. Of course, there is quite a bit of restriction at those holes, so only a certain

amount of fluid can pass at a time (this is the resistance to movement you feel when trying to move the shock by hand). This disc arrangement is the valving that controls the flow of oil. If more force is applied to the rod, the discs deflect farther away from the holes in the piston, allowing more oil to pass through. The stronger the discs, the greater resistance to oil flow, hence why one shock will have more resistance to motion than another.

Any restriction to normal flow, whether it's hydraulic oil or electrons in an electrical circuit, creates heat. So as oil passes through the valve in a shock, it heats up. The faster the oil

An inside view of a remote reservoir shock, in this case Bilstein's 5150 series. Note the Bilstein version is fixed to the shock body, while other designs connect the reservoir to the main body of the shock via a flexible braided steel hose. (Courtesy Bilstein USA)

capable of exerting on the motion of the suspension. This is more commonly known as shock fade, and is the reason why you can sometimes feel the shocks become less effective during long sections of high-speed running down a rough road.

One way to reduce aeration is to put fluid under pressure, which you might recall from science class raises the boiling point of a fluid (this is why an engine's cooling system is pressurized). Hence, pressurizing the cylinder of a shock raises its boiling point and increases resistance to shock fade.

Though all shocks operate on the same basic principles just described, there are three major types of shocks available for automotive applications.

Twin-Tube

By far the most common shock available from the aftermarket, most basic replacement shocks sold with lift systems are twin-tube hydraulic shocks. With this design an inner cylinder contains the oil, piston, and rod. A second chamber wrapped around the outside holds extra fluid that is circulated through the main cylinder. This style is a good all-around shock for everyday use and it has extra oil capacity over OE versions to help reduce shock fade. There are also gas-pressurized twin-tube shocks, which are a step above standard shocks in terms of quality and can better hold up to high-speed use. Generally speaking, the charge on a twin-tube gas shock is lower than other shock types.

Monotube

As the name implies, there is only one cylinder in this type of shock that contains all of the oil and internal components. Monotube

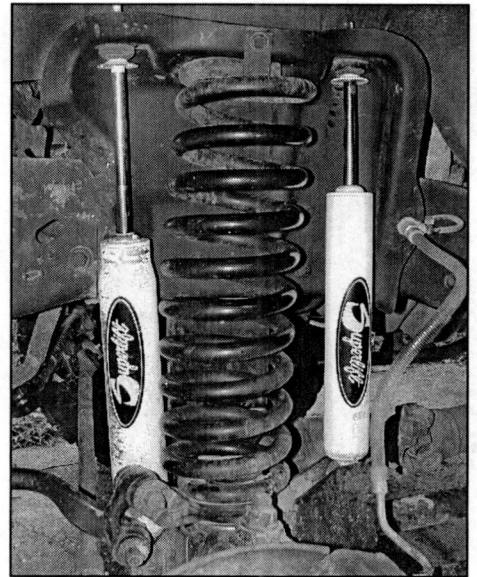

Twin-tube hydraulic shocks are the workhorses of the aftermarket with thousands sold every year. In most cases they are an upgrade over the factory shocks and are better suited to handling the rigors of off-road use. For this reason, installing aftermarket shocks is a good idea even if the suspension is left stock.

shocks typically have a high-pressure gas charge to help prevent aeration. In the case of Bilstein monotube shocks, this gas charge is separated from the shock oil by a floating piston to prevent the gas from mixing with oil. High-pressure monotubes tend to be more responsive and precisely valved for a particular application. Monotubes are also used in applications where space is tight and a wider twin-tube body just won't fit. Monotube shocks are an excellent choice for coil-spring applications where the shock needs to exert a great deal of control over compression and rebound. High-pressure monotubes are also typically used as the main body of a strut in late-model 4x4s and high-performance aftermarket coil-over shocks. As stand-alone

is forced to move through the shock valve, the more heat gets built up. If you touched a shock on a vehicle after running down a rough road at speed (something that is not recommended), you would be amazed at how hot a shock absorber can get! The shock bleeds off this heat through the metal walls of the cylinder, which are usually subjected to airflow, as the vehicle is in motion to help it cool off. If the oil gets hot enough, however, tiny air bubbles start forming in the oil, which is called aeration. These tiny air bubbles can pass though the valve more quickly than oil, which reduces the amount of resistance the shock is

A part of the off-road racing world for many years, remote reservoir shocks provide extra cooling and oil capacity to help withstand the tremendous beating that racing dishes out. As with many things, this racing technology has trickled down to light trucks.

Remote reservoirs are primarily for looks enhancement on most production vehicles, but to many the extra hassle of mounting the reservoir is well worth it.

Thankfully the trend in the '80s of adding as many shocks per wheel as possible has passed. Dual shock systems are beneficial for most applications, while triples are arguably more for looks than functional enhancement.

components, these shocks are sometimes included with standard lift systems, but are more frequently seen as optional upgrades.

Remote Reservoir

Up until the past few years, remote reservoir shocks were only seen in the racing world, but this technology has since trickled down to the production lift kits. As the name implies, this design has an external reservoir attached to the main body of the shock by either a braided stainless steel hose or piggy-backed directly to the outside of the shock. This reservoir contains extra shock fluid that is circulated through the main body of the shock as the shock cycles. This external reservoir provides the shock with extra fluid capacity and a place for the oil to bleed off heat before it is returned to the main body. Remote reservoirs have a high gas (usually nitrogen)

charge and a monotube main body since all of the extra fluid is contained in the external reservoir. These high-quality shocks are usually rebuildable and a select few can even be re-valved by someone who really knows what they're doing. Though they provide the truck with an unmistakably "racy" look, reservoir shocks have very little practical purpose on most trucks. Unless lots of high-speed off-roading is in the future, most real reservoir shock benefits come from high-quality construction and precision valving rather than the presence of the external reservoir. Looks count for a lot these days, but be prepared to pay a premium as these shocks are expensive to manufacture.

Multi-shock Systems

Another trend with truck enthusiasts is to add one or two extra shocks per wheel, and several lift manufacturers enable truck owners to do just that via bolt-on systems. Multi-shocks are beneficial because

the dampening duties are spread across more than a single shock, so each one doesn't have to work as hard (and heat up as fast) to effectively control the suspension. However, multi-shock systems can also create a firmer ride because adding shocks increases the resistance to motion unless the shocks are valved to work in a dual or triple arrangement. Again, this upgrade largely boils down to enhancing looks, as many systems add a tubular hoop that is visible in the wheel well for that "race-inspired" look. Multi-shock systems are truly beneficial for high-speed environments and with coil-spring systems where additional shock valving is needed.

SOLID FRONT AXLE WITH LEAF SPRINGS

This Chapter Includes:

1941–1995 Jeep CJ/YJ
1969–1987 Chevy Pickup
1969–1991 Chevy Blazer/
　　　　　Suburban
1969–1993 Dodge truck
1999–2004 Ford F-250/F-350
　　　　　Super Duty
1966–1977 Toyota Land Cruiser
1979–1985 Toyota Pickup
Various other years of Ford trucks,
International trucks, full-size Jeeps,
and other domestic and import
vehicles

There are many instances in the mechanical world where the original design becomes the standard by which all others are judged, and the solid axle/leaf spring suspension system is an excellent example. It is not a coincidence that the list of vehicles that utilized this design is a "who's who" of off-road legends. With few exceptions, every truck is a well-respected and sought after vehicle for off-road enthusiasts. Solid axles and leaf springs have traversed the deserts and jungles of the world, won the Baja 1000, and delivered

Solid-axle Chevys are never a bad choice when building a trail rig. These vehicles are plentiful and cheap, but, not so coincidentally, one that has not been cut up and beat to death is becoming hard to find. But when it comes to simplicity and great drivetrain combinations, '73–'87 Chevys are hard to beat.

critical people, supplies, and material under fire.

Simplicity has a lot to do with why this system is so popular. There are relatively few moving parts, and fewer parts mean fewer chances for

something to go wrong. It is also versatile; a leaf spring suspension can be made to carry just about any load imaginable over any possible terrain. It is easy to modify; because of its simplicity and versatility it can

The originator and still king to many, the old Jeep flatfenders (or flatties as most off-roaders affectionately call them) were some of the first 4x4s to be subjected to suspension modifications. Even today they are still a popular choice as evidenced by this photo taken less than two years ago. Even though coils are increasingly replacing leaf springs, it's hard to deny a leaf-spring suspension's simplicity for all but the most aggressive terrain.

be modified without a degree in mechanical engineering. But perhaps the ace in the hole is its performance potential; with minor modifications, a leaf spring suspension system can be made to perform well in any environment. But before it can be modified, it's important to understand how it works.

Basic Anatomy

All solid-axle/leaf spring suspension systems are basically the same. Leaf springs are attached to the frame via spring bushings. At the center of the springs, U-bolts attach the springs to pads on the axlehousing. Because springs can serve as locating members for the axles and support vehicle weight, there are no other locating points needed. A sway bar attaches to each side of the axle and the frame to control the amount of "sway," or body roll, which

improves cornering. When in four-wheel drive, power passes from the front driveshaft to the differential and then out through two-piece axleshafts to the front wheels.

Of course, this basic design has a

few variations. Later-model vehicles, such as the Ford Super Duty and Jeep YJ, also include a track bar (also called a panhard rod). This bar takes over locating the axle side to side. A track bar enhances the handling of the vehicle. Some vehicles, such as the Toyota trucks, also incorporate a traction device for the front axle. This track arm, not to be confused with a track bar, is mounted parallel with the front springs and prevents the front axle from wrapping under load (see more information about axle wrap in Chapter 8).

In all cases, lifting a leaf spring suspension involves installing leaf springs with more arch than the original pieces. Because the axle is one piece, the taller springs simply move the axle farther away from the frame, causing lift. There is no need for extensive cutting or welding, and most of the time the lift can be done by a competent shade tree mechanic, provided the proper tools are available. But as discussed in Chapter 1, there is much more to a lift than a pair of springs.

The leaf spring design in modern interpretation—other than placing the springs on top of the axle rather than below and the addition of a track bar, not a whole lot has changed from the original designs dating back to the '40s.

TYPICAL FORD "T" STYLE STEERING DETAIL

DRAG LINK   PITMAN ARM

TRACK BAR BRACKET

TRACK BAR

TIE-ROD   EXTREME RING

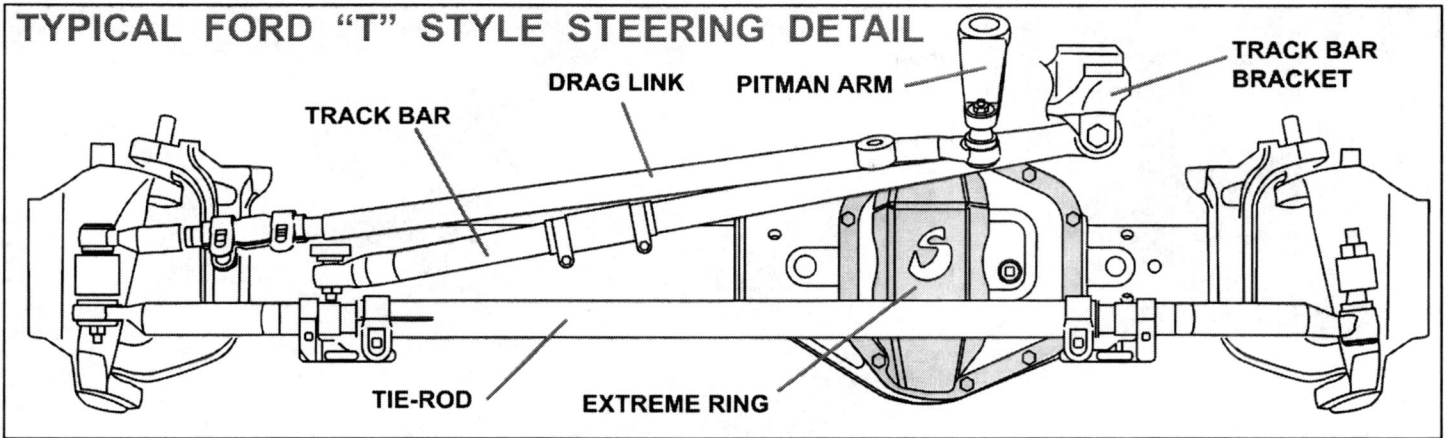

Traditional inverted "T" linkage. This design is much more receptive to a lift than other linkage styles, which is why some systems are converted to this type when a lift is in the equation. (Courtesy Superlift)

Steering

The biggest differences among the solid-axle/leaf spring crowd have to do with the different steering systems used. All of the steering systems are adequate, but some are more susceptible to lifts than others.

Inverted "T" Linkage

The most common type of steering system is an inverted "T," which is used on Jeeps, Fords, and other vehicles. A pitman arm is attached to the steering box and pushes/pulls a drag link. This pitman arm moves parallel to the front axle. The drag link in turn either attaches directly to the passenger side steering knuckle or to the tie rod close to the knuckle on the passenger side. The tie rod connects the two front tires together. As the vehicle is lifted, the angle of the drag link changes and gets steeper. With minor amounts of lift, this change is minimal and no modifications are needed. At about 4 inches of lift (sometimes less depending on the vehicle), the angle becomes excessive. The most common correction is to replace the factory pitman arm with a "drop"

pitman arm. A drop pitman arm has more distance between where it attaches to the steering box and the drag link. This reduces the angle of the drag link thus bringing it back into its factory operating specifications (which is always the goal as discussed in Chapter 1).

Drag link angle is always important, but it becomes critical on vehicles equipped with a track bar, such as Super Duty trucks and YJs. On these vehicles it is important for the drag link to have the same arc of movement as the track bar. If the two are out of phase with one another, bumpsteer and steering wheel kick usually result. For this reason, the lift system should always include some form of *both* steering *and* track bar correction, not just one or the other. Track bar correction usually happens via brackets that either lower the bar's attachment point on the frame or raise it at the axle.

"Chevy Style" Linkage

Though this type of linkage does not have a commonly used name, this same basic system was used on Dodge and Toyota trucks in addition to Chevys. With this arrangement,

With Chevy- and Dodge-type steering linkage, the drag link is short and runs parallel to the frame. The drag link should operate roughly level to the ground, and the goal with a lift system is to keep these operating angles as close to factory as possible. Though it's a little quirky, this style of steering linkage can be corrected using common off-the-shelf parts with up to 6 inches of lift. Beyond that, a crossover steering system is the best choice for anything but show trucks.

the pitman arm on the steering box moves perpendicular to the front axle. A short drag link connects to the pitman arm and a steering arm on top of the driver-side knuckle. A tie rod connects the driver and passenger knuckles together.

Chevy steering in real life is pretty simple to understand. This truck is equipped with a 6-inch lift and has both a 4-inch raised steering arm on the knuckle and a 2-inch drop drag link that runs between the steering arm and the pitman arm (the pitman arm and steering box are just out of view above the top of this photo).

Straight axle Chevy trucks are one of the few models that appeal to both the off-road enthusiasts and the show crowd. Lift heights range from 2 to 12 inches of lift for this era, but many agree a truck with 4 to 6 inches of lift with 33- or 35-inch tires strikes the perfect balance of utility and looks.

While this system may be perfectly serviceable in stock form, it becomes a little problematic when lifted much over a couple of inches. This is due in large part to the drag link; because it is so short, its operating angle increases dramatically with minor amounts of lift. Correcting this type of linkage takes one of three forms. The first is to replace the factory drag link with a "drop" drag link, or an "S" shaped link that allows the drag link ends to operate without binding. Another form of correction involves installing a raised steering arm. This arm has a raised drag link attachment point, once again to relieve operating angle. A less desirable version of the raised steering arm is a steering block, in which a machined piece is installed under the factory steering arm, thus spacing it upward. The third is specific to Chevys and involves a "drop" pitman arm. Any, or a combination, of these methods works adequately up to about 8 inches of lift. But for those wanting to go over that lift height or wanting to use the vehicle heavily off-road, many experts recommend replacing this entire system with a "crossover" steering system. Crossover steering essentially converts the vehicle to the inverted "T" system, but extensive modifications are needed to accomplish the conversion. In most cases a different steering box is needed along with fabricated bracketry to mount the box to the frame. Fortunately, conversion kits are available for popular applications, including Chevy and Toyota trucks.

The Specifics

Even though there is a wide range of vehicles that fall into the "leaf-spring solid axle" category, it is inevitable that there are variations and eccentricities among the various manufacturers that utilized this suspension design. And because they remain very popular among enthusiasts, what follows are some tips and tricks specific to the more popular models.

'69–'91 Chevy trucks

Though the Jeep may have been the mass-produced originator of the leaf-spring design, Chevy brought it to the masses on a huge scale—literally. Other full-size trucks may have utilized leaf springs prior to Chevy's introduction in 1969, but Chevy brought the design to period perfection and new levels of popularity. These trucks, combined with the ½-ton Fords and Jeeps of the same era, birthed the suspension industry as we know it today. As a result, the sky is the limit on what you can get for these trucks. Note that the technical year split is '69–'87 Chevy pickup trucks and '69–'91 Blazers and Suburbans.

The Good

Virtually everything about these trucks is good. There are the legendary engines and transmissions put out by Chevy during the era (some better than others of course) and the 12-bolt rear/Dana 44 front axles of

Though all straight-axle Chevys are desirable, various equipment changes make some year models more desirable than others for the off-road crowd. The 10-bolt front axle on this '82 is not as attractive to off-roaders as the earlier trucks that used Dana 44s, and the 10-bolt rear axle on this same truck is inferior to the earlier 12-bolt. Some of these trucks used full-time transfer cases that are a little quirky (though part-time conversions are available) and the later ones are often plagued with anemic 305 V-8 engines. Thankfully, installing beefier axles, engines, and transmissions from other trucks of the same era (including ¾- and 1-ton variants) is often as simple as bolting them in place.

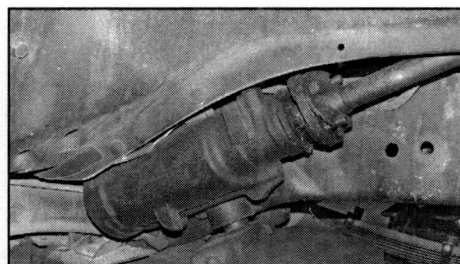

The steering box mounts on these trucks are notorious for cracking under the added stress of larger tires. In fact, the frame around the steering box should be very closely inspected prior to purchasing one of these vehicles. The tell-tales are stress cracks that run between the box mounting holes and the box "rocking" on the frame when turning. Any cracking or movement here at all should be fixed immediately because it will cause all sorts of problems. Reinforcement plates are readily available and should be considered mandatory when running 35-inch or larger tires.

the early years. Due to their enormous popularity, nearly all pieces of these vehicles are available as replacement parts. Suspension components are no exception. Because just about every company manufactures a lift for these trucks, the market is very competitively priced, making these arguably the least expensive vehicles to lift. They are dirt-simple vehicles to work on, and the suspension does not get much more straightforward. As far as a trail vehicle is concerned, look for the '87 pickup and '87–'91 Blazers and Suburbans with fuel injection despite their less desirable axles. Second-best is '72–'78 vehicles for their Dana 44 front axles. Suspension-wise, they are practically unchanged.

The Bad

The steering can be problematic on these trucks, especially on taller lifts, even though correction parts are readily available. The issue, often as not, has to do with age. The newest truck from this era is now almost 20 years old, and most have had an owner at some point in their past that has been lackluster about maintenance. These days one should expect the typical wear on ball joints, steering components, and suspension parts. Some or all of these components may need to be replaced in addition to a suspension lift in order to make it run straight down the road.

One special note about these vehicles is that they are well known for developing frame cracks around the steering box, especially with the prolonged strain of larger tires. Those that intend to use the vehicle for serious off-road use or who want to run a lift taller than 6 inches should seriously consider reinforcing the frame

at the steering box mounting location with inexpensive and readily available reinforcement kits. The ultimate fix for hard use or extremely tall lifts is a crossover steering system. Though these systems require extensive modifications, including the installation of a steering box from a two-wheel drive and fabricating the mounts to go with it, these systems eliminate the short drag link and the problems that go with it.

What to Look for in a Lift System

As mentioned before, the lift systems on these vehicles are extremely price-competitive, sometimes to the point that quoted prices do not include all of the components necessary. One common trick is to omit shocks from the advertised price, even though replacement shocks are needed with as little as 2 inches of lift. Also be sure that steering correction is included, and pay attention to what type. Most steering correction methods are acceptable, but a steering block is the least desirable of these because they place quite a bit of strain on the studs that attach the steering arm to the knuckle. These should be avoided because there are other methods that perform the necessary correction more effectively that are not much more expensive.

Pay attention to other accessories on the vehicle that may also need to be addressed. One such example is sway bars. Some of these vehicles came with one, while others did not. As a result, sway bar correction is almost always listed as an accessory even though having a sway bar is beneficial for on-road driving, and not relocating the sway bar can negatively impact its effectiveness and overall vehicle ride quality. Brake hoses are another beneficial upgrade. Even though reputable lift systems include some sort of brake hose relocation if needed with a particular lift height, due to the age of these vehicles, the brake hoses are probably marginal and could stand replacing anyway.

Other age-related upgrades to consider are replacement shackle bushings and sway bar bushings. It is very rare to see front or rear shackle bushings included with a standard lift system, and these bushing can have a profound impact on ride quality if they are excessively worn. The same can be said of sway-bar bushings.

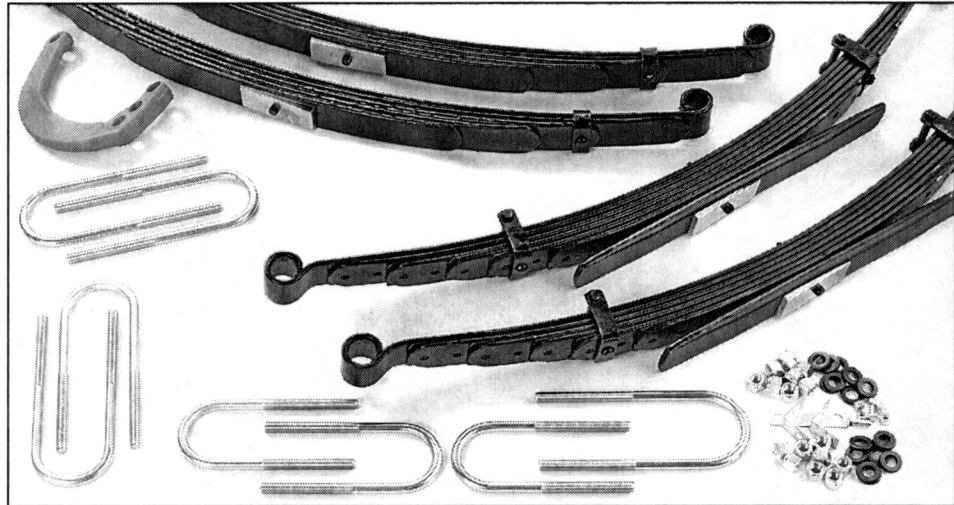

Lift components vary according to lift height. The 6-inch system shown includes all the basics as well as the rear springs, which are the most desirable option, but blocks are also acceptable for the rear. This kit also includes a steering arm, which is always better than a steering block and just as good as a drag link. This kit also includes replacement U-bolts, which are needed on any truck that ranges from 20 to 35 years old. The other kit is a 12-inch system, and the needs are more complex. Rear springs are mandatory, and all three forms of steering correction are needed to adequately address the factory steering design. Still, a crossover system is a better option, and most likely both front and rear driveshafts need to be lengthened. Also, do you see shocks anywhere in either of these photos? (Courtesy Superlift)

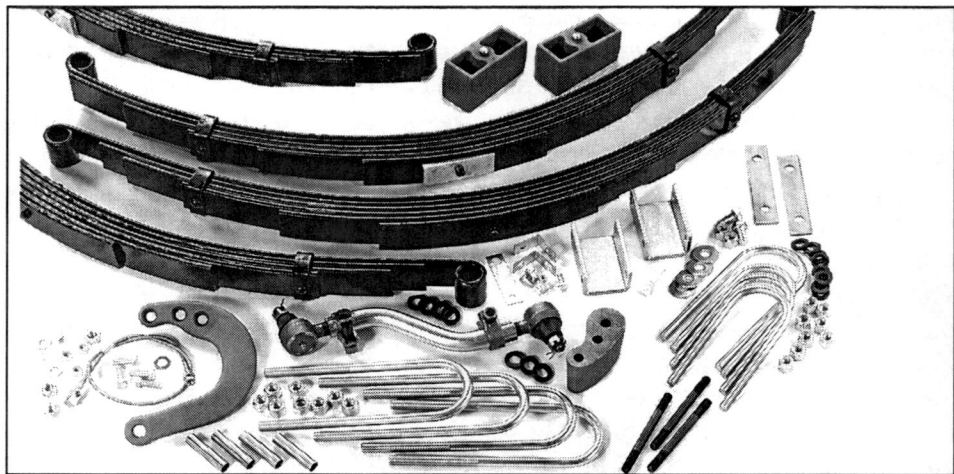

What Fits, What Hits — Chevy/GMC

Model	Year	Modification	Tire Size (inches)								
			31	32	33	34	35	36	37	38	40
½, ¾, & 1 Ton	1969–'91	None		1½	2½	4	6	8		10	12
Leaf Spring		Fender Trim			1½	2½	4	6		8	10

Lift kits always include new front spring bushings, but that means only half of the shackle bushings are addressed. It's very rare to see new shackle bushings for the frame included in a straight-axle Chevy lift, but considering the age of these trucks, spending a little extra on new ones (which are readily available) along with the minimal time needed to replace them is a wise choice.

Though panned early on due to Jeep's sacrilegious deviation from traditional round headlights and plagued with some highly failure-prone components in its first years, the YJ Wrangler is a highly respected vehicle in off-road circles, even though some people will admit this only grudgingly. (Courtesy Skyjacker)

Perhaps the biggest decision to make regarding these lift systems has to do with the method of rear lift. The pros and cons to the various rear lift methods are discussed elsewhere, but it should be noted that the factory rear springs are most likely sagged with age and use, so as a result using rear lift blocks may yield less lift than is needed to maintain a level stance. This makes rear springs often the best option for these vehicles and a must for anyone that uses the truck off road. Note that with few exceptions, prices for moderate lift kits are quoted with blocks for the rear.

'41–'95 Jeep CJs and YJs

The venerable Jeep has been the subject of more books about 4x4s than any other, so as a result a multitude of resources are out there (and corresponding different schools of thought) to help Jeep owners modify all aspects of these off-road legends. From engine swaps to drivetrain alterations to suspension modifications, you can find it all. There is even a strong movement for radical suspension mods to these vehicles that involve replacing the leaf springs entirely with coils and coil-overs. Chapter 9 touches on these high-performance systems, so for now let's stick to the basic tried-and-true lift methods.

The Good

Jeeps are small, nimble, cheap, and readily available, so one cannot really go wrong choosing a Jeep as the foundation for an off-roader. For the most part they come with reliable (if usually underpowered) engines, satisfactory transmissions, and acceptable axles. Of course, there are several

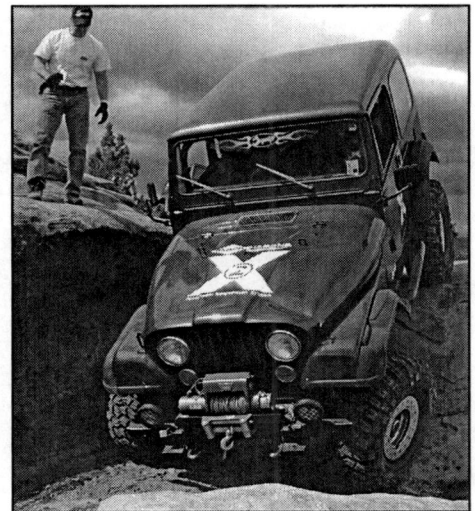

Jeepers are a fiercely loyal and adventurous bunch, and owning a Jeep makes you the automatic member of an exclusive club. A leaf-spring Jeep's off-road prowess is legendary and making just a few modifications to the suspension and axles can allow you much greater ability to explore the backcountry.

Short of lowering the transfer case's position to relieve driveline angle, lifts in the 4- to 6-inch range usually require a driveshaft equipped with a constant velocity (CV) joint. On most CJs this is a very simple modification, while YJs require an SYE kit that is discussed more in Chapter 8. The rear driveshaft is usually the most problematic, but the front can also be a problem at taller heights.

Jeep CJs suffer from notoriously weak frames. Though usually only a problem when subjected to severe use, the main issue is that the frames are "C" shaped and not fully boxed. Many Jeep builders box the frames themselves, though this requires a lot of time and extensive fabrication skills. Later YJs have much stronger fully boxed frame rails.

notable exceptions in each category, but that is a subject best left for dedicated Jeep books. As for the suspension, it did not change much throughout the decades, with most changes corresponding with the introduction of new models. Because of their popularity, suspension lifts are readily available and inexpensive.

The Bad

Their attribute as a short vehicle is also the source of a common problem: driveline vibration. The short wheelbase leads to short rear driveshafts, and those short driveshafts mean driveline angles increase much faster as ride height increases than their longer wheelbase brethren. Throwing the virtually unlimited number of drivetrain and axle combinations possible in a Jeep only aggravates the issue. For this reason, lift systems stay relatively mild, usu-

ally in the 4-inch range. Of course, it is possible to go higher (and people often do), but taller lifts usually lead to very expensive modifications involving hybrid drivelines, custom axles, and stretched wheelbases. If you plan to go sky-high and build a show truck, there are better choices out there.

The early Jeeps suffer from weak frames and a highly undesirable (not to mention overly complicated) steering system known for its ability to wear out quickly. The best remedy for these vehicles ('41–'71) is to throw away the entire system in favor of a Saginaw conversion. Though common, the conversion is expensive and usually involves welding. As for the frames, they are well known for cracking and bending under hard use, which can cause an uneven stance (leaning), not to mention an unsafe vehicle. Look carefully

CJ-2As, 3As, and 3Bs all utilize "C" shaped shackles that are not popular among off-roaders. Though greasable, these shackles are not as strong as the more traditional shackles of today. CJ-5s and 7s use better shackles, but they are constructed of thin stamped steel and are also known to fail when subjected to severe use. Fortunately, heavy-duty replacements are available for both shackle styles.

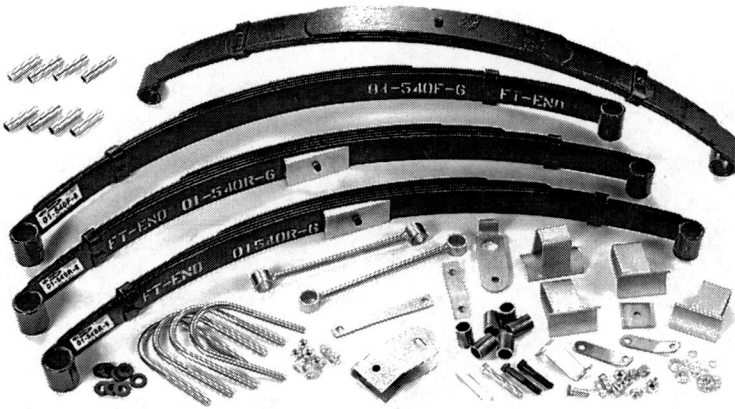

A quality Jeep lift kit includes front and rear springs, U-bolts, compression stop extensions, sway-bar links, transfer case lowering spacers (not usually needed with a CV driveshaft), brake hose extensions, and shocks. This YJ kit also includes brackets for the front and rear track bar. (Courtesy Superlift)

A Jeep equipped with an aftermarket dropped pitman arm. Note that drop arms are available for both manual and power steering boxes, but the two are not interchangeable.

for cracking, particularly around the spring and shackle hangers. Later Jeeps do not suffer from these problems and have notably stronger frames, but they have other issues, mostly drivetrain and axle related.

One theme common to all of the Jeeps covered in the year split is weak spring shackles. The very early Jeeps had failure-prone C-shaped shackles, while virtually all of the others were composed of fairly thin stamped steel. The shackles are also short, which can cause spring eye-to-frame interference when the suspension flexes in some cases. All of these applications can benefit from stronger aftermarket shackles, especially those with a brace in the middle, but beware of those substantially longer than stock, as discussed below.

What to Look for in a Lift System

Jeeps are "spring under," meaning that the axles are attached to the top of the leaf springs rather than the bottom like most trucks. As a result, lift blocks and other inexpensive lift methods used on trucks will not work on a Jeep. A proper Jeep lift system includes at minimum a set of front and rear leaf springs, U-bolts, brake line relocation brackets, and shocks. Of course, competing price points sometimes list these items as "optional" even when they're not, so buyer beware.

Beyond the basics, there are other items listed as options that should be seriously considered, mostly due to the age of these vehicles. Brake lines are always a good choice, as are shackle bushings. The front shackle bushings in particular can cause all kinds of scary handling traits if they are worn out. The usual complement of other items should be checked and replaced as necessary, including tie rod ends, wheel bearings, ball joints, and U-joints. Ensuring the vehicle is in tip-top condition increases the chances of being satisfied with the lift system.

Steering correction is one area in which there are two schools of thought. For Jeeps 1975 and earlier, there are not many options, which is why lift is usually limited to 4 inches or lower. On '76 and newer models, most of the lift manufacturers recommend using a drop pitman arm at 3½ inches of lift and above. How-

ever, many off-road shop owners and installers recommend against a pitman arm at 3½ to 4 inches of lift on CJ-5s, CJ-7s, and YJs. They site reduction in the steering box's mechanical advantage, bumpsteer, and reduced turning radius among the reasons to stick with the factory pitman arm. This argument tends to be further bolstered by a few lift manufacturers not automatically including a drop pitman arm with their 4-inch systems, even though they recommend their use. Therefore, steering correction may be more a matter of personal preference, or at minimum dictated by how an individual vehicle is equipped. Virtually all of the experts agree a pitman arm is needed above 4 inches.

The Good and Bad of Other Modifications

Aside from the basic lift systems, there are innumerable other suspension accessories on the market that can seem like a real boon for off-roaders. Here is a closer look at some of the more popular accessories out there and what they impact.

increase suspension articulation off-road. And they do exactly that: disabling the sway bars on a vehicle can substantially increase its performance in slow-speed off-road situations where lots of suspension travel is needed to negotiate rough obstacles. Disconnects are a much better alternative to removing the sway bars from the vehicle entirely, which can hamper the Jeep's on-road stability. Most disconnects are inexpensive and safe when used properly, making them worthwhile additions for any off-roader. However, these are not a good choice for high-speed off-road use.

Extended-length shackles: Replacement shackles by themselves are not a bad idea as most are much stronger than the factory pieces. However, some extended-length shackles on the market are advertised to add lift, usually 1 to 3 inches. Remember that a shackle only lifts one half of the spring, usually netting only half of the claimed lift. Also, extended shackles should be avoided on the front of these vehicles because they reduce caster, which causes wandering and all sorts of other weird handling quirks. Used on the rear, however, extended shackles can improve driveline angle. Generally speaking, extended shackles beyond 1 inch should be avoided.

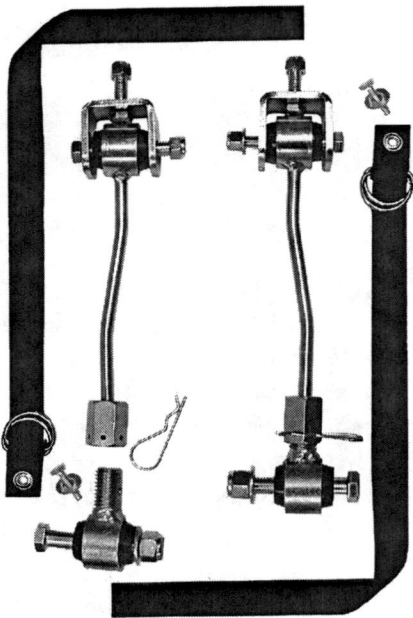

Sway bar quick-disconnects are replacement links that have an easy way to disconnect the sway bar for slow-speed off-road use. Many different disconnect designs are out there, but they all revolve around the idea of being able to defeat the sway bar without using tools. These particular disconnects are made by Superlift, but many other lift manufacturers offer them as well. (Courtesy Superlift)

Heavy-duty aftermarket shackles are always a good upgrade, but avoid installing shackles that are significantly longer than stock in order to gain lift. Shackles much more than 1 inch over stock length can cause several different problems that are not readily apparent. This shackle is a little longer to accommodate the military wrap on the spring, but also note it is boomerang-shaped to aid in rear bumper clearance. Boomerang shackles are a good choice for YJs with aftermarket bumpers.

Sway-bar quick-disconnects: Though there are many, many different kinds of quick disconnects out there, they all do basically the same thing: they give the operator an easy way to disconnect the sway bar(s) in order to

What Fits, What Hits — Jeep

Model	Year	Additional Modification	Tire Size (inches)								
			31	32	33	34	35	36	37	38	40
CJ2A, CJ3A, CJ3B	1946–'63	None	1	2½							
		Body Lift 2"		1	2½						
CJ5, CJ6, M38A1	1955–'86	None	1	2½	4						
		Body Lift 2"		1	2½		4				
J-10/20	1974–'90	None					4				
Wrangler YJ	1987–'96	None	1½		3½						
		Body Lift 2"			1½		3½				

Spring-over conversions involve relocating the springs to the top of the axles (factory position is below). They have many benefits, including better ground clearance and generally more suspension travel. However, it takes a lot more work than many people realize to make a spring-over function properly.

A CV-style rear driveshaft can often eliminate the need to lower the transfer case, which also reduces ground clearance. CV driveshafts are nearly mandatory on short-wheelbase Jeeps that are used off-road.

"Goofy" shackles: All manner of articulating, sliding, and other types of "goofy" shackles can be found if one looks hard enough. Regardless of the design, "goofy" shackles are designed to increase the movement of the suspension, reduce bind, and generally improve off-road performance. The trouble with most is that they increase off-road performance at the profound sacrifice of stability. Furthermore, adding the shackles often causes other problems, such as driveline bind, axle wrap, and steering issues. Bottom line: buyer and fabricator beware. Often these seemingly simple devices can cause more trouble than benefit.

Spring-over conversion: A seemingly simple modification, a spring-over involves moving the Jeep's axles from the top of the springs to the bottom (like most trucks). A fairly simple modification to perform (simply weld four spring perches above the originals), spring-overs net 4 to 5 inches of lift and flex well because the relatively flat stock springs are re-used and more leverage is placed on them. However, there are a number of factors that must be addressed in order for the spring-over to perform properly, most notably the steering and some sort of device to control axle wrap. With the steering, the springs are in the way of the tie rod and the lift over-extends the factory linkage. And since the springs have much more leverage on them than they were designed to take, driveline-destroying axle wrap is common. All of the issues a spring-over creates can and have been tackled successfully, resulting in a well-performing suspension system when done properly. But there are so many variables involved that the proper steps to take are largely vehicle specific, and addressing them all takes skill and experience. In other words, stay away from this modification unless you have the skills, patience, and budget to make it come out right.

Avoiding the transmission crossmember drop: At 3½ inches and above, most lift systems provide spacers to lower the transmission crossmember, which in turn relieves driveline angle, but at the detriment of ground clearance. Furthermore, many people find the spacers unsightly. On CJs, this can often be avoided by replacing the factory rear driveshaft with one equipped with a constant-velocity (CV) joint. A CV shaft can handle more angularity than a conventional driveshaft, so the transmission crossmember can remain in the stock location. On YJs, installing a slip-yoke eliminator kit is required in order to run a CV driveshaft. Slip-yoke eliminator kits replace the factory output shaft in the transfer case and have the added benefit of increasing driveshaft length and preventing transfer case fluid loss if the rear driveshaft breaks. Though not required for a lifted vehicle, these modifications are beneficial for off-road use.

'99–'04 Super Duty Trucks

A testament to just how popular and durable the solid-axle/leaf spring design is, Ford returned to this configuration when it introduced its redesigned heavy-duty truck line in 1999. Thanks in part to this return,

Many an off-road enthusiast applauded Ford's return to traditional solid axles and leaf springs with the new line of Super Duty trucks in 1999 (actually select mid '90s Fords were leaf-sprung solid axles too, but this was limited to some 1-tons). An instant hit with 4x4 enthusiasts, they quickly proved to be equally as capable of venturing off the pavement as they were in more traditional roles as work trucks. Then again, is this really a surprise?

Super Duty trucks went through three different U-bolt changes: square (shown), round, and "large radius" in which the U-bolt makes a total of four bends to complete its "U." Ford wasn't very consistent about when they made some of these changes, so taking a peek under your Super Duty to verify it's U-bolt style may very well save some frustration when installing a lift.

the Super Duty was an instant hit. Perhaps the ultimate work/play truck, they are strong enough to haul just about anything, powerful enough to do so quickly, and easy enough to modify for just about any purpose. All of these attributes make the Super Duty a great choice for someone looking for a combination work/play vehicle, provided tight trail use isn't in the cards.

The Good

This generation of Ford enjoys strong support from the aftermarket, so just about anything is possible. Even though they are late-model vehicles that benefit from the latest in technology, they are exceedingly receptive to modifications, be it suspension or drivetrain related. Though the vehicle's ride may not be as good as the later models with coil springs, it is acceptable to most provided suspension lift choices are done wisely. Even though these vehicles sport an "antiquated" design, they corner well on the street while providing decent

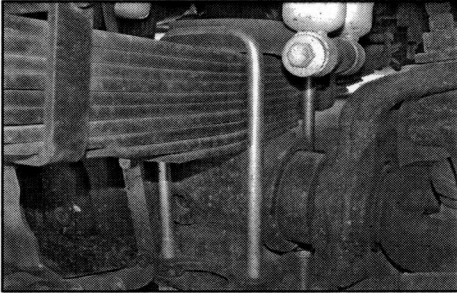

This generation of Super Duty trucks utilizes a factory block with a large tab on it that serves as the pad for the compression stop. As a result, the factory blocks must be retained in order to keep the compression stops functional. Unfortunately, this means stacking blocks when a rear lift block option is chosen, which is never recommended for off-road use. Tall blocks also tend to induce axle wrap.

performance and flex off the highway. Best of all, no platform since the solid-axle Chevy is easier and more receptive to taller lift systems. So if it's a show truck you want, the choices could not get much better.

The Bad

Not much can be jotted down in the "bad" category. The early '99 trucks had a rather obscure track bar change that can play havoc when determining exactly which lift system is needed for the vehicle. They also went through three different front U-bolt changes and two sway bar link modifications, and in typical Ford fashion, it can be difficult to determine exactly when those

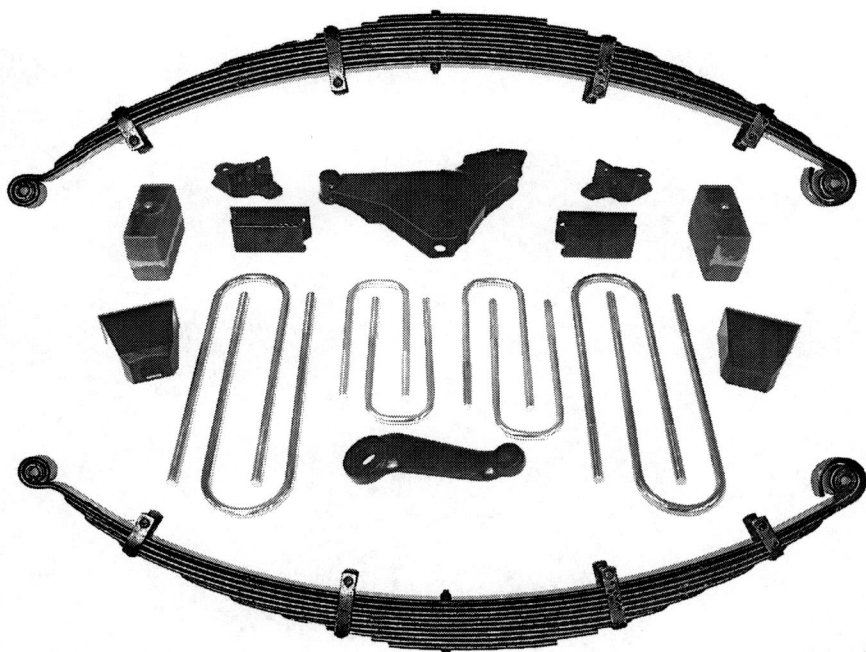

A typical 4- to 6-inch lift system includes new front springs, U-bolts, a track bar bracket, a pitman arm (6 inches only), compression stop extensions, sway bar brackets or links (depending on year split), and shocks. Since several manufacturers offer different shock options for these trucks, it is wise to do some research on which shocks will work best for your intended use. (Courtesy Superlift)

Track bar correction is important on a Super Duty, even one with leaf springs. Short 2-inch systems still include a new track bar bracket for the frame in most cases. Fortunately the factory bracket is bolted in place rather than welded or riveted, so installing the new bracket is a remove-and-replace procedure. Note this particular bracket is used on both 4- and 6-inch systems, as there are two possible mounting holes for the track bar itself. Since this truck has a 4-inch lift (as evidenced by the factory pitman arm), the bar is mounted in the upper hole.

changes were made. Examine the application guides carefully to be sure you order the correct parts.

These trucks utilize a track bar, which by itself is not a bad thing because a track bar enhances the vehicle's on-road stability. However, if the track bar and drag link do not remain in the same arc of movement, severe bumpsteer will result. These trucks are very sensitive to changes in the operating angle of these two components, and if they are not "just right" some very unsatisfactory handling characteristics will result.

Another attribute that also has a bad side is the tremendous amount of torque the Power Stroke diesel engines put out. All that torque can cause axle wrap and driveline vibration if lift blocks are added to the rear. Lift companies counteract this by a variety of different means, some of which are better than others. Eliminating the lift blocks entirely and installing rear springs is one of the most effective means, but this reduces load-carrying capacity quite a bit. At the installer level, some shops recommend replacing the two-piece driveshafts commonly in these trucks with a conventional one-piece 'shaft, but this is an expensive option. Just be aware that lifting diesel trucks at best may result in a little launch shudder (brief vibration when accelerating from a stop) and at worst may require traction bars, rear springs, a new driveshaft, or a combination of the three.

What to Look for in a Lift System

First and foremost, these vehicles were sold with both gas and diesel

Dual-shock kits are a popular option for these trucks, which can benefit from the extra dampening due to their heft. As for the shocks themselves, standard hydraulics (shown here) are good, as are monotube gas and remote-reservoir shocks.

engines, and the diesels are substantially heavier. As a result, be sure you go with a company that offers gas- or diesel-specific springs. This indicates they have built the substantial weight differential into their spring's design. Using diesel springs on a gas truck severely degrades ride quality, while gas springs on a diesel truck do not maintain lift height and sag prematurely under the additional weight. Using springs designed for the application ensures a more satisfactory finished product. As far as spring design, some companies have opted to apply thin multi-leaf technology to these trucks in an effort to optimize ride quality, while others have stuck with conventional 5- to 8-leaf packs. It can be argued in both directions which design is better, so it is best to ask around and then make an educated decision about which is better.

As mentioned above, it is important for a lift system to address both drag link and track bar angle, even with as little as 2 inches of lift. As a result, better lift systems include track-bar correction specific to each lift height, usually via a bracket at the frame that replaces the original piece. Taller lift systems may also include an adjustable track bar to help ensure the axle stays properly centered under the vehicle.

The vast majority of the lift systems currently on the market for these vehicles include just about

Solid-axle Toyotas are wildly popular among the moderate-to-hardcore off-road crowd because they are plentiful, cheap, and well supported by the aftermarket. These attributes, plus the exceptionally well-designed drivetrain, make solid-axle Toys a great foundation for building a dedicated trail machine.

everything you need to lift properly, so looking at some of the other accessories offered is a good plan of attack. Premium shocks, multi-shock systems, skid plates, dual steering stabilizers, brake hoses, and traction bars are all worthwhile upgrades for these vehicles. Some of these accessories might even be included in a system as a value-added incentive or special promotion. The bottom line: keep an eye on the ads for special promotions on these popular trucks.

'79–'85 Toyota Pickups

In these days where 4x4s are a dime a dozen it's hard to remember that four-wheel-drive mini-pickups have not been around for very long. Toyota jumped into the market along with everyone else in the late '70s, but while most of the competitor trucks from this era have long been residents of scrap yards, these Toyotas are still alive and plentiful. Not only are Toyotas amazingly reli-

What Fits, What Hits — Ford											
Model	Year	Additional Modification	Tire Size (inches)								
			31	32	33	34	35	36	37	38	40
F-250	1999–'04	None				2	4	6		8	10
F-350	1999–'04	None				2	4	6		8	10
Excursion	2000–'03	None				3		5			

One of the big reasons Toyotas are popular among the trail-going crowd has to do with the transfer case. Several years ago a small company called Marlin Crawler developed a way to mate two transfer cases together, thus multiplying the truck's crawl ratio exponentially. These days many companies support these trucks with heavy-duty axles, low gearsets, and body armor. Drop-in 4:1 kits are now available for these 'cases along with twin-sticks and dual or even triple 'cases, so the sky is the limit on slow-trail crawling.

able, they also perform exceedingly well off the pavement. Once some key products hit the market to increase the low-range ratio of these vehicles, their fate in the off-road world was secured. Just like a Jeep, it's hard to go wrong selecting one of these trucks as the foundation for an off-roader. But most importantly, what sets this era of Toyotas apart from later versions is that simple and venerable solid front axle.

The Good

Even though these trucks are fairly old, parts and accessories are still readily available. The 22R four-cylinder engine that powers the majority of these trucks will run forever (some say they even thrive on neglect), and the rest of the drivetrain shares the same reliability. Their diminutive size makes them nimble in off-road situations, and because they are old mini-pickups, solid-axle Toys can be found dirt cheap. Suspension-wise they are equally simple, with the same reliability evident that is in the rest of the truck. Perhaps best of all, there are several off-road shops and small manufacturers that

make Toyotas their specialty, so high-performance off-road suspension and drivetrain parts can be found for just about any application.

The Bad

They are not getting any younger. Even the best-designed vehicles eventually succumb to wear and tear, and while 200,000-mile Toyotas are common, it also means a good chunk of its useful life is behind it. The engines, while respected for their longevity, are also world-renowned for being underpowered, especially for a deceptively heavy truck. One of the most controversial parts of these trucks is also one of the big reasons they're famous: the solid front axle. Unlike virtually all other modern front axles, Toyotas use a "closed knuckle" axle and Birfield joints rather than conventional U-joints. While more than adequate for stock trucks, these axles become the weak links in the system once lower gears

Solid-axle Toys seem to thrive on abuse. No matter how biased someone may be to a particular brand, everyone has to admit that Toyotas can take a beating and keep on ticking.

One popular modification to consider when building a trail rig is to replace the oddly designed factory steering with a crossover system like the one shown here. A crossover conversion requires installing a steering box from a later IFS truck along with steering arms for each knuckle. These offer better control when the suspension flexes and places the tie rod above the springs, out of harm's way.

Rear lift should be chosen carefully, especially for a trail rig. Replacement springs are usually the best way to gain the necessary amount of lift out back since the factory springs have most likely sagged out.

Toyota front axles are a "closed knuckle" design in which the pivot points for the knuckles are enclosed inside a giant ball on either end of the axlehousing. Very old domestic axles were built the same way. Inside this ball is a Birfield joint rather than a traditional U-joint; Birfields are known to be problematic, and it's rare to go to a Toyota event without seeing someone fixing a broken one. Aftermarket upgrades are available to significantly beef up the relatively weak factory parts. Stock Birfields can be a problem with 35-inch tires, low gears, and lockers.

upgrade parts available from the aftermarket.

More directly related to the suspension, the steering system also leaves a lot to be desired. Similar to a Chevy, a short drag link running perpendicular to the front axle transfers input from the steering box to the steering arm on the driver side knuckle. Making matters worse, the drag link uses an antiquated method of connecting to the pitman and steering arms. Like the Chevys, the best method of correction is completely replacing it with a crossover steering system. While kits are readily available, this upgrade is expensive.

What to Look for in a Lift System

The standard warnings about the lift kit in question being complete

and larger tires are thrown into the mix. Some dedicated off-roaders replace these axles with other units considered stronger, while others swear by the design and upgrade the factory axle with a wide range of

still apply. Front springs, U-bolts, shocks, rear lift, and steering correction are all included in a complete lift system. The most common method of steering correction is an "S" shaped drag link that uses factory-style ends. This is the preferred method because the factory drag link usually has a substantial amount of wear anyway, provided a crossover system is not being considered. Select lift manufacturers may also offer a raised steering arm that replaces the factory arm on the driver's side. However, both methods are borderline if the vehicle is going to be used off road.

Spring design should be closely considered when shopping. Even though lift systems are common for these trucks, many of them are built using decades-old prints from a time when ride quality was much less of a consideration than it is today. For some of the larger manufacturers, it's simply not practical to go back and redesign springs for an older application whose volume is low compared to newer vehicles. At the same time, companies that make Toyotas their specialty can focus more attention to these trucks and often have lift springs manufactured to their specifications. As a result, these are the applications where it may be wise to give specialty companies more

emphasis, even if they are more expensive. As discussed elsewhere, it is not wise to judge a spring simply by the way it looks, so doing some research and then making an educated decision based on individual priorities is advantageous.

One last thing to consider is rear lift. By far the most common method is lift blocks, as only a select few companies offer replacement rear springs. However, blocks are not the best choice, as it is very common to see rear springs that have sagged and deformed with age; in fact, it is unusual to find a Toyota from the era that has serviceable rear springs. Therefore, spending the extra money on rear springs is a big plus, especially for off-road use. Some specialty companies even offer rear springs that are much longer than stock. Though this requires some fabrication to the spring mounts, the longer spring provides a smoother ride and more articulation than a stock replacement lift spring.

Other Applications

Even though each make and model has its own little quirks that should be known to the owner before making modifications, practically all of the common quirks with leaf spring/solid axle trucks have

"The rest" of the solid-axle leaf-sprung crowd includes Suzuki Samurais, full-size Jeep trucks, and early Dodge trucks like this Ramcharger, among others. Most platforms have a limited amount of products available from the aftermarket, but the thing to remember is it's hard to go wrong with leaf springs at all four corners regardless of the vehicle. (Courtesy Skyjacker)

been addressed in the four categories above. Fortunately there is a specialty website for just about any vehicle, whether it is popular or obscure, where owners can share knowledge and experience with others. While any advice on these sites should be taken with a grain of salt (there are far more people on the Internet that think they're experts than actually are), generally speaking these can be useful resources for specific information on these and other vehicles.

What Fits, What Hits — Toyota											
Model	Year	Additional Modification	Tire Size (inches)								
			31	32	33	34	35	36	37	38	40
Pickup	1979–'85	None		3	4		7				
		Fender Trim			3	4		7			
4-Runner	1984–'85	None		3	4		7				
		Fender Trim			3	4		7			
FJ-40	1964–'80	None	2½								
		Cut-Out Flare			2½						

Boosting a YJ

Jeep owners have been lifting their vehicles since the first civilian models began rolling off the assembly line because we all figured out early on that more ground clearance and more tire equals more off-road prowess. Even though not a whole lot changed in the Jeep world suspension-wise until 1997, lifting a leaf-sprung short wheelbase CJ or YJ properly requires a delicate balance of driveline and steering correction as inches are added. And it is also due to their short stature that not much more than 4 inches can be had without extensive and expensive modifications.

To take a closer look at what's involved with lifting a Jeep properly, we installed a Black Diamond 3½-inch lift system on an otherwise stone stock YJ. Black Diamond's system is representative of what is found with most name-brand lifts in that everything you need to do the job properly is included, so there are no hidden costs involved. The main differences among the lift systems is the design and rate of the springs, which impact's longevity of the lift system, but more importantly dictates how the Jeep is going to ride after the deed is done. The Black Diamond springs appear to have all the right stuff, with tapered ends, military-wrapped eyes for durability, anti-friction pads, and polyurethane bushings. The spring rate, however, is a compromise between topless YJs and those with added hardtop weight. This is a

"one-size-fits-all" mentality that not all lift manufacturers share, but time has shown the Black Diamond springs to have the ride and handling down along with the ability to hold up when the Jeep is loaded down with gear.

The lift installation is pretty much the same for CJs and YJs. The procedure is fairly basic when compared to full-size IFS stuff, but it can be daunting to beginners. The tools and equipment needed are fairly simple, but a couple of special tools are required that we'll point out as we go along. This installation was done at a shop with a vehicle hoist, but the job can be done in a driveway by a competent shade tree mechanic and some friends, which brings up a point: having an extra set of hands is always a good idea when installing a lift. The extra cost of pizza and beer is well worth it. Barring any unforeseen delays like frozen bolts, two guys can tackle the job in one long day or an easy weekend. Not bad for substantial improvements in both looks and off-road performance!

The Black Diamond kit proved to be an affordable and easy way to accommodate 33x12.50-15 Pro Comp Mud Terrains. The sum of these parts has transformed a dirt-cheap, basically stock YJ from a commuter into an eye-catching weekend toy equally at home on the beach, in the dunes, or on the trails.

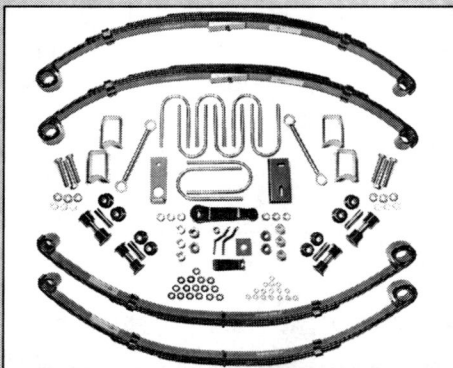

1. All of the standard Black Diamond lift-system components are here. The kit includes front and rear leaf springs, U-bolts, shock absorbers, transfer-case lowering spacers, bumpstop extensions, sway-bar links, front and rear track bar relocation brackets, brake hose relocation brackets, bushings, and all of the necessary hardware along with easy-to-understand, well-illustrated instructions geared towards shade tree mechanics. All Black Diamond components are backed by a limited lifetime warranty.

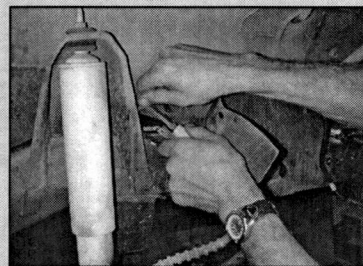

2. Pretty much any lift beyond 2 inches is going to require addressing brake hose length in some shape or form, and YJs are no different. Because the axles are 3.5 inches farther from the frame, either the stock hard line must be carefully extended or longer hoses must be used. Nothing is wrong with re-using the original hoses in a different location, and it is cheaper than purchasing new hoses. However, if budget permits, go ahead and spring for the replacement hoses. They look cleaner and are more accommodating to off-road suspension travel, especially once the sway bar is disconnected. When bending/extending the OE lines, be careful not to kink the tubing. (Tom Morr)

Boosting a YJ *continued*

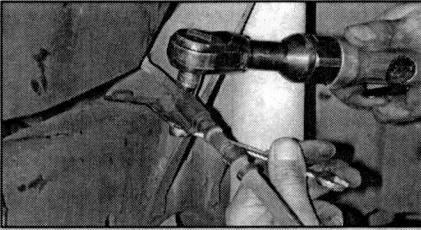

3. Replacement hoses were not in the budget this time around, so the installers made use of the Black Diamond zinc-plated brake brackets for the front. These secure to the frame at the stock hold-down location, then extend the hard-line-to-hose fitting out and down so that the factory hose can handle the added suspension droop. Remember, new brake hoses can always be added later.

4. Necessary specialty puller tools can often be borrowed from your local auto-parts store. However, most guys use a project like this as an excuse to add a few items to the toolbox. A pickle fork is an inexpensive but handy device for separating tapered steering joints. Tie rod ends are not a pickle fork's only talents, however. In this case, the sway-bar links were proving to be stubborn, so the assistance of a pickle fork allowed the links to see the error of their ways and cooperate by separating from the sway bar.

5. The introduction of square headlights also saw the addition of a front track bar. The Jeep engineers added track bars to enhance the vehicle's handling and cornering ability over that of the CJ. Like the steering and other components, track bar operating angle is changed as ride height increases. Therefore, the kit includes a simple front track bar bracket to ensure that the axle stays centered.

6. Polyurethane is great bushing material, but it will drive anyone insane with incessant squeaking if it's not lubricated. Always add a liberal dose of grease when installing poly bushings in anything. Opinions vary on which is best, but any good moisture-resistant grease works. In this case white lithium grease minimizes squeaks and helps the steel sleeves slide into the Black Diamond polyurethane spring bushings.

7. This Jeep was virtually stock, but at some point in the past someone installed some extended spring shackles. Since the stock ones were nowhere to be found, there was no choice but to re-use the extended shackles. These are about the maximum length that should be used with a lift system, adding about an inch to the overall equation. Predictably, however, the rear shackles over-corrected the driveline angle and there was some vibration after the lift was complete as a result. Installing the springs is a remove-and-replace procedure both front and rear. The springs do install a certain way, and they're clearly marked FRONT for idiot-resistant installation.

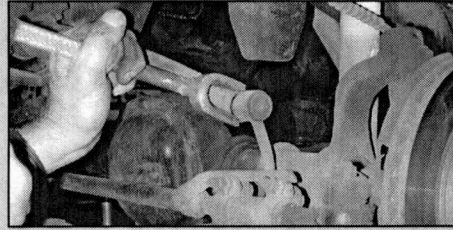

8. New U-bolts should always be included with a lift system, especially with older vehicles. The kit includes bump stop extension brackets that are positioned on top of the axle tubes and which install under the new U-bolts. These engage the rubber bump stops on the frame when necessary and keep the springs from over-compressing if the Jeep ever gets airborne (which it shouldn't, but you never know). Not so coincidentally, they also keep the tires from bashing the fenderwells.

Boosting a YJ *continued*

9. A proper lift system handles all of the details, no matter how small. This one includes extended sway-bar links that enable the bar to do its job with the lift system. Installation just requires lubricating the supplied bushings, installing them in the links, and then bolting the links in place.

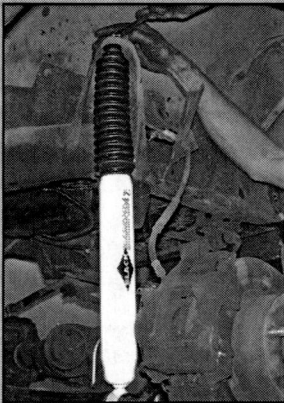

10. It is a well-known fact that a sway bar hampers slow-speed off-road performance, so one option that Black Diamond offers is a quick-disconnect sway-bar link kit. A pull-pin allows the upper piece to unscrew and be strapped to the sway bar, thus allowing full suspension droop. Sway bar quick-disconnects come in an astonishing variety of designs, but no matter what kind they are a worthy upgrade for a trail rig. Like the brake hoses, quick-disconnects can be added later when budget permits.

11. As the condition of the hands indicates, at this stage of the game you will realize that suspension work is a very dirty business. Shock options abound for YJs. Value-minded systems like this kit include standard hydraulic shocks, in this case lightly valved Black Diamond ATs. Gas-charged units are a viable option, but realistically remote reservoirs are not needed on a leaf-spring Jeep unless racing is in the cards.

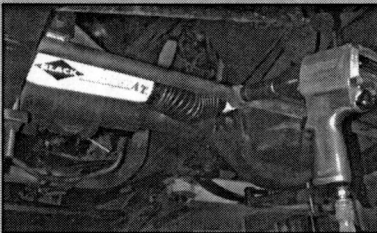

12. Here is the bone of contention among experts on lift systems: some consider a dropped pitman arm mandatory at this lift height, while others say it's not necessary. Black Diamond "recommends" using a dropped pitman arm, but it is nonetheless an optional item with the lift. Other systems include the pitman arm. Generally speaking, it's best to go with what the manufacturer recommends. Installing the arm requires using a puller tool to remove the old one; these can usually be rented at the local parts store. The drag link can be separated with a pickle fork, but use caution to avoid damaging its rubber boot.

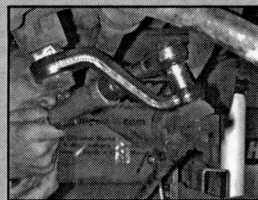

13. Steering stabilizers are almost always optional, but are beneficial because they help absorb road irregularities and keep the Jeep tracking properly. Even if the Jeep has a stock stabilizer, it's usually small and not up to the task of controlling larger tires and wheels. Quality-manufactured single cylinders are usually adequate for 35-inch tires, and the factory location (shown here) is usually the least susceptible spot for the stabilizer to incur off-road damage.

14. The stock steering linkage can be retained with a lift system, which makes sourcing replacement parts as easy as walking into the local parts house. Closely inspect all of the drag link and tie-rod ends for wear, as the spooky steering characteristics caused by marginal steering parts is greatly amplified with a lift system. Always replace any suspect parts. You can also see how the dropped pitman arm lowers the operating angle of the drag link, keeping the drag link ends from binding and as the suspension cycles.

Boosting a YJ *continued*

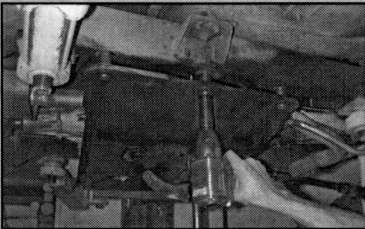

15. Lowering the transmission crossmember (or belly pan) is a necessary evil with short-wheelbase Jeeps. Spacing down the pan also lowers the transfer case, which in turn relieves driveshaft operating-angles to avoid vibration. At 3 to 4 inches this is mandatory when using the factory driveshafts. The lowered pan reduces ground clearance, which is a big no-no for a trail rig. However, the pan can remain in its stock location by installing a slip-yoke eliminator kit in the transfer case. An SYE kit adds length to the rear driveshaft and adds a CV joint capable of working at a greater angle than conventional U-joints (see Chapter 8).

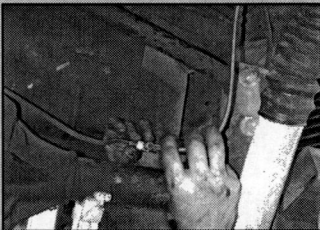

17. As mentioned earlier, a harder-core option is to upgrade to extended-length brake hoses. If the replacement hoses are used, the relocation brackets supplied with the kit can be discarded. However, the brake system must be opened and later bled to remove any air pockets. Always use line wrenches on brake fittings, which are prone to stripping out.

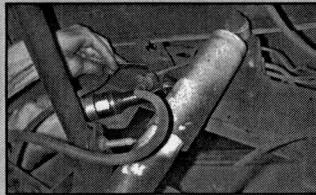

16. Just like the front brake hoses, the rear hose must either be replaced with a longer one or the factory line must be relocated downward. Black Diamond supplies a simple bracket that bolts to the frame, and then the metal line is carefully re-formed to reach the new lower mounting point.

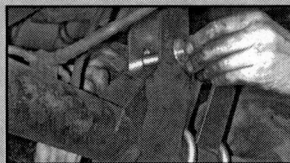

18. Another variant from the CJ models is a rear track bar, which aids in stability. Like the front track bar, the mounting point at the axle is simply raised and the stock bar is retained. Many trail-only YJ owners discard the front and rear track bars, as they are more or less not necessary for slow-speed maneuvers. It should be emphasized that this can only be done if the intended use is off-roading exclusively.

19. After bleeding the brakes and double-checking all fasteners for proper torque, the installer bolted on a set of 33-12.50-15 Pro Comp MTs and went for a spin. Right away the enhanced ground clearance was evident, as the Jeep dropped into a hole that would have left a stock Jeep teeter tottering on its frame rails. The suspension also flexed surprisingly well, but ultimately open differentials and the wrong line left the Jeep unable to pop back out the other side without the aid of a winch. Lockers anyone?

SOLID AXLE WITH COIL SPRINGS

This Chapter Includes:

'66–'79 Ford Bronco
'73–'79 Ford F-150
'05–current Ford F-250/F-350
    Super Duty
'97–current Jeep Wrangler (TJ and
    JK)
'84–'01 Jeep Cherokee
'94–current Dodge Ram

Even though coil-sprung 4x4s started in the mid '60s, the design is still popular today and is used under a variety of light-duty as well as heavy-duty trucks. (Courtesy Skyjacker)

If a solid axle/leaf spring suspension is the gold standard, solid axle/coil spring suspension is the silver. Around almost as long as the leaf spring system, the coil system shares many of the same attributes: reliable, easily modifiable, and can be adapted to perform well in just about any environment.

Coil springs are attractive to suspension designers because they are compact, inexpensive to manufacture, and generally offer superior ride characteristics when compared to an average leaf spring. There is no friction in a coil spring and they can even be made to have a progressive spring rate. However, a coil spring has none of the structural attributes that a leaf spring does, so there must be a series of links or attachment points to properly locate the axle. Because a coil spring is not a structural member of the suspension there are more components involved in locating the axle, which makes a coil suspension more complicated. Even so, with ride quality being the major emphasis of suspension design these days, suspension designers gladly put up with the increased complexity. This is why there are several different coil-sprung vehicles manufactured today.

Designers have used a variety of different methods to locate the axle, and this book examines each in detail. Because of the increased complexity, of course, more things can potentially go wrong. This is why it's

Even though substantial technology improvements have been made in virtually all areas of the current truck crop (especially compared to 20 years ago), coil spring suspension has remained largely the same. This simple, time-honored design is also easy to lift, and lift higher. This 2005 Super Duty is equipped with an 8-inch kit and can clear 38s with no trouble.

Many coil-spring trucks are equipped with single shocks at each corner (the factory position is inside the coil spring on this Dodge). However, single shocks are often not up to the task of controlling a heavy-duty coil suspension. As a result, several lift manufacturers offer multi-shock systems that can add one or two more shocks per wheel. Though the need is questionable on a street-bound truck, it doesn't hurt. Not so coincidentally, multi-shock systems also look pretty cool. (Courtesy Skyjacker)

important to be sure you have everything you need to lift the vehicle properly. Before diving into brand-specifics, there are a few areas common to all coil spring suspension systems that bear mentioning here:

Shocking Matters

While coil springs have all of the attributes mentioned above, they have no internal friction. This means that all of the spring rate is built into the heavy-gauge wire that makes up the coil. On one hand this is an attribute because it is easier to "tune" a coil for desired ride or performance characteristics, but on the other hand this also means a coil can release its energy as fast as it is introduced. Just like compression, if that release is not controlled it can mean very negative ride quality. Since shocks are the only component with the job of controlling the introduction and release of energy, they are an important consideration in a coil spring design because they must be strong enough to control the springs without introducing negative ride characteristics themselves. Shocks with improper valving can cause an overly "bouncy" ride or just the opposite: a stiff, bone-jarring ride. A shock also tends to get worked harder with a coil spring, which is why it is not uncommon to see dual front shocks from the factory and/or a pressurized gas shock that is capable of reacting more quickly to the constant introduction and release of energy. Because coil springs are more

difficult for shocks to control, investing in premium shocks and multi-shock systems is a good idea, especially if your intended off-roading involves high-speed runs.

In a Bind

Chapter 2 includes an explanation of the different schools of thought on coil spring design and construction. While there may not be a clear advantage of one design over another, the real message should be to not judge a coil spring strictly by its appearance. What is important, however, is that coil spring bind should *never* happen with a coil spring suspension. Coil bind is when the coil spring is compressed to the point that all of the

Compression stop position is critical with a coil-spring system. If the stops are left in the factory location, the coil risks compression to a point where it binds against itself. This 6-inch lift includes both a drop bracket and a progressive urethane stop to prevent such occurrences.

individual wraps are forced together. This does two things: 1) It causes a bone-jarring spike as the force that compressed the coil is now transferred directly to the passenger compartment in high-speed situations, and 2) It causes major coil fatigue and premature failure by either sagging or fracturing. Coil bind is not exclusive to high-speed maneuvers. It can also happen any time the suspension is forced to fully compress (such as when rockcrawling). The best way to prevent coil bind is to ensure that the compression travel stops, or bumpstops, have been addressed with the lift system. This is done by either relocating the factory bumpstops or by using longer replacement pieces. If high-speed desert or sand running is in the plans for the truck, investing in a quality set of progressive-rate bumpstops is a good way to ensure a better ride in the rough and longer coil spring life.

The Infamous "Death Wobble"

Though any suspension can do it, coil spring systems seem to be more susceptible to "death wobble" than other types. Death wobble is a common suspension industry term that describes when the front tires and/or the steering wheel begin shaking uncontrollably at slow-to-moderate speeds, and the only way to stop it is to stop the vehicle. Not only is death wobble very dangerous (it can cause complete loss of vehicle control), it is really scary for the person behind the wheel. Why coil spring systems seem to be more susceptible than other types is mostly a mystery, but a strong case can be made for the number of individual components needed to locate the axle.

Death wobble can rear its ugly head during a variety of situations. The most common is going over an irregularity in the road, such as a pothole, with just one tire, but accelerating, turning a corner, or even going over a dip in the road can also start it with no warning. There are two important things to remember with death wobble: 1) If it happens, stop the vehicle as quickly and safely as possible because steering input rarely makes a difference, and 2) Once it happens, it happens again until the source is found.

So what causes it? The answer is either a little slack in a few key components or improper alignment. The latter is easy to fix provided the lift system (if equipped) has been designed properly, otherwise the answer is not so simple. Look very closely for play in the ball joints, wheel bearings, steering linkage, steering box, steering shaft, suspen-

The steering shaft (the shaft that connects the steering column to the steering box) is an often-overlooked source of play in the steering system. Some vehicles are known to have poor and wear-prone designs that can introduce a surprising amount of slack between the column and shaft. U-joint-equipped shafts are generally good, but rubber rag joints and bell-shaped joints are not. Fortunately, high-quality aftermarket replacement shafts are available for most of the problematic applications, but they can be expensive.

sion bushings, track bar, and even shock bushings. Tires with irregular wear, that are heavily out of balance, or using tires with bias ply construction on a vehicle designed around radials can cause it. Bent suspension/steering components and even bent axlehousings have been known to create death wobble. Keep in mind there may not be one clear culprit; sometimes all it takes is a little wear in two or more components to cause it. If you are unfortunate enough to experience death wobble, you should focus all of your attention to fixing it as quickly as possible.

Coils vs. Coil-overs

One of the recently emerging trends among the popular vehicle platforms in this category (namely

It's hard to go wrong with the high-quality coil springs available from suspension manufacturers these days. They are designed with a delicate balance of ride quality, load-carrying capacity, and the ability to maintain proper ride height for a long period of time. Simply put, conventional coils are more than adequate for both street trucks and off-roaders.

Race-inspired coil-over conversions are the latest trend in high-performance lift systems. These coil-over shocks from Skyjacker are designed for a late-model Dodge and open up a whole other realm of off-road possibilities. Though coil-over upgrades come at a premium over conventional coils, it's hard to deny they set the standard for looks. Best of all, these coil-overs come with bracketry to bolt them in the place of coil springs, so there's not a bunch of torch work involved to put them on. (Courtesy Skyjacker)

TJs, late-model Ford trucks, and Dodge trucks) is the development of high-performance systems that replace the coil springs and shocks with a coil-over shock. Long a staple in the racing world due to their compact construction and virtually unlimited tune-ability, coil-overs are the gold standard in the off-road market. Though coil-overs have been adapted to Jeeps for years, the limited availability of heavy-duty coil-overs and their extraordinary cost has made them an extreme rarity on heavy trucks outside the desert-racing world until recent years. Fabtech

and Skyjacker have both released largely bolt-in systems to convert Dodge and Ford trucks to coil-over. Though expensive, these conversions take the off-road performance of these trucks to a whole new level: Imagine bombing down a washboard road at freeway speeds in a ¾-ton truck and the picture begins to emerge. These conversions ratchet up the "cool" factor to new heights as well. While the off-road performance gains are undeniable and the ride is unsurpassed, they can impact load-carrying capacity and on-road handling. If the goal is building a dedicated pre-runner, it is hard to beat these systems and their "out-of-the-box" performance. However, the costs may be difficult to justify for a

vehicle that is depended upon to serve in more traditional roles.

Ford Trucks

Ford's version of coil spring suspension debuted in 1966 with the introduction of what is now a legend: the Ford Bronco. And even though this version has always had its quirks, this same basic design continues on the current (2005 and newer) Super Dutys. It consists of two radius arms that locate the axle front-to-rear. These arms attach to the axle above and below its centerline, so they also control the "roll" of the axle, or caster. A track bar handles side-to-side axle location. Overall it is a fairly basic system that

The original mass-produced coil-spring 4x4 was the Ford Bronco. First released in 1966, the Bronco was an instant hit and didn't change for 11 years. These days early Broncos have reached cult status and are still extremely popular among off-road enthusiasts. Not surprisingly, many aftermarket suspension parts are also available.

works well as long as it is properly maintained.

Lift System Basics

Because of the system's relative simplicity, lift kits are fairly basic as well. They usually consist of longer coil springs, shocks, and some sort of caster correction. It is with caster correction that the differences begin.

Radius Arm Matters

The early trucks utilize a radius arm that sandwiches a roughly diamond-shaped pad cast or welded to the front axle. In between the two is a two-piece C-shaped bushing to absorb road vibrations and give the suspension some movement. In stock form the radius arms are parallel with the frame, but as lift

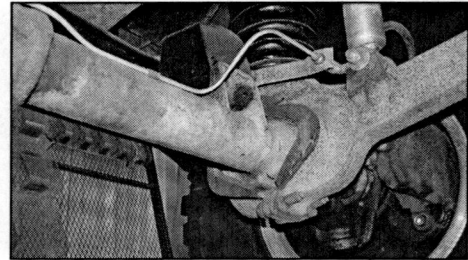

Though it disappeared for a few years, Ford's radius arm/coil-spring suspension design continues today with the introduction of the late-model Super Duty. First released in 2005, coil-spring Super Dutys are top sellers. (Courtesy Superlift)

The early Bronco radius arms are physically clamped to a pyramid-shaped piece on the axlehousing, a rather odd design that hasn't been seen before or since. Sandwiched in between the housing and the arm is a C-bushing. Though this worked surprisingly well, the C-bushings are known to deform, split, and work their way out of the radius arm connection over time. The original rubber bushings have long since rotted in most cases, but thankfully urethane replacements are readily available. Offset versions are also available that reposition the axle within the radius arm to restore the caster lost when adding a lift.

increases, the radius arms become angled because the axle end moves downward as the frame end remains fixed. As the arms move from parallel they also "roll" the axle and reduce caster. To restore the lost caster, lift companies provide new C-bushings with a certain amount of offset to "roll" the axle back to its proper location. These bushings are available in 2-, 4-, and 7-degree offsets. A simple and effective way to address caster, the replacement C-bushings are a good idea for another reason: the trucks that use them are at least 30 years old and replacing the wear-prone C-bushings is a good idea.

Another correction method that is applied to both old and newer trucks lowers the radius arm attachment point at the frame. This brings

Radius arm drop brackets are the most common form of caster correction whether it's an old truck like the one pictured or a late-model Super Duty. This replacement bracket lowers the radius arm attachment point, keeping the radius arm more or less parallel to the ground just as it was stock. The early versions often required torching the original brackets off of the frame, but thankfully the late-model trucks have a bolt-in bracket that just requires drilling a couple of holes.

the radius arms back to more or less parallel with the frame and thus restores caster. While just as effective as the C-bushings, one drawback to this method is a reduction in ground clearance. Some taller lift systems use a combination of the two methods, but this is really only necessary with lifts above 6 inches.

A third method of caster correction comes in the form of replacement radius arms. Though this is the most expensive option, these arms typically have the necessary caster correction built in. Also, most are longer than stock, which enhances suspension travel (a longer lever on the axle allows it to move farther and more freely before suspension components bind). Typically made of high-quality materials, these replacement radius arms offer a strength advantage over the factory parts and have the added bonus of a "racy" look.

Expanding on the replacement radius arm theme, the late-model trucks can also be upgraded by replacing the radius arm entirely and converting to a four-link system similar to what is found on Dodge

trucks. Most of these conversions bolt to or very near the factory radius arm attachment point at the frame, while select ones are actually longer than stock. The real on- and off-road benefits of a four-link conversion over a replacement radius arm are debatable, but there is no denying that tubular links look far better than a lowered factory radius arm.

When addressing the radius arms on a Ford, don't overlook the bushings at the frame. The '60s and '70s trucks used a bayonet style bushing that, while offering much in the way of movement, also has a reputation for wearing out. Strangely, virtually all of the lift systems do not include new radius arm bushings as standard equipment, but these should be considered mandatory replacement items regardless. The newer trucks use conventional steel-encased rubber bushings.

Steering

For the most part, the steering systems on these trucks are simple and effective. A traditional steering box is connected to the steering

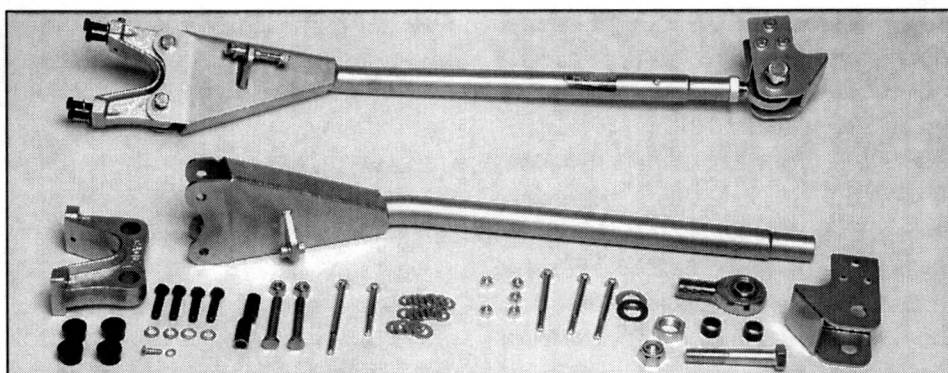

At the high-end of the spectrum are replacement radius arms such as these from Cage Offroad. These tubular arms offer substantial suspension travel improvements and the required caster correction to work at greater ride heights than stock. These even do away with the factory bayonet bushings in favor of a massive heim joint at the frame. These are available for both early Broncos and the larger solid-axle pickups and Broncos.

Four-link conversions are commonly available for late-model Super Dutys and replace the bulky factory radius arm with a decidedly racier four-link conversion. Caster correction is built right into the design of the arms and they improve suspension articulation by reducing the amount of bind the factory arms exhibit. Four-link conversions are usually available as an upgrade over conventional lift systems. (Courtesy Skyjacker)

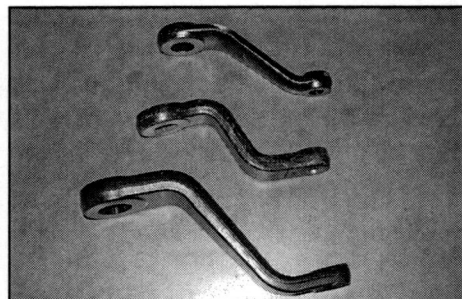

Pitman arms come in a variety of shapes and sizes, but fortunately coil-sprung Fords are pretty consistent and have very clear year splits when changes were made. Still, it's always a good idea to verify pitman arm part numbers (as well as the rest of the lift components) to be sure you have received the proper parts. This simple process can avoid later delays and vehicle down time.

The factory bayonet bushings have a mixed reputation in the off-road world. They allow quite a bit of flex before they bind up but they also tend to wear more quickly than conventional bushings. This is one case where urethane bushings tend to wear better and last longer than the rubber bushings they replace. Be sure to purchase new bushings when ordering a lift for an older truck, and pay attention to how the old bushings come apart; how they go together can be confusing and installing the new parts incorrectly will destroy the bushing in a short period of time.

knuckles via an inverted "T" linkage. The one exception is the '76 and '77 models equipped with power steering; these used an inverted "Y" linkage system that does not respond well to lift systems at all. Fortunately, correcting the problem is as simple as replacing all of the linkage with the components found on the later '78–'79 models. Even though this modification is only needed at 6 inches of lift, it should be seriously considered regardless of lift height.

In all cases, steering correction comes in the form of a drop pitman arm. On some of the older vehicles, it may also be necessary to install an adjustable drag link in order to restore full turning radius. Once again, due to the age of these vehicles, this is not a bad idea regardless.

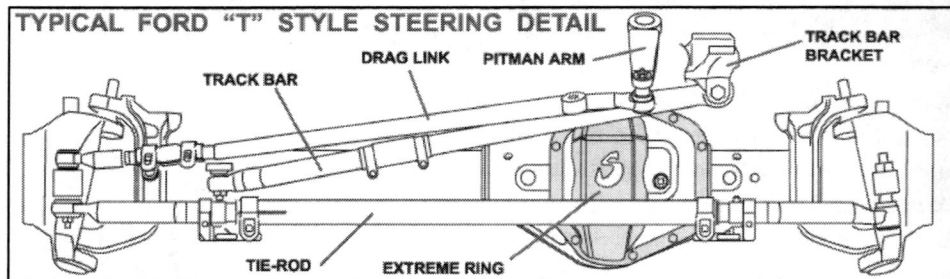

Most early Fords come straight from the factory with the inverted "T" style linkage shown. However, the '76–'77 models have a funky inverted "Y" type linkage in which the drag link meets the tie rod about two thirds of the way to the passenger side steering knuckle rather than going all the way to the knuckle as shown. The "Y" shaped linkage is not at all conducive to lift systems. If your truck is one of the few years that has the inverted "Y" linkage, converting to the style shown will be necessary when lifting the truck.

A typical adjustable track bar for early Fords. In some cases these aftermarket pieces are cheaper than stock replacements, and the aftermarket versions are normally rebuildable, so it's the last track bar you'll ever have to buy.

Early Ford kits can consist of as little as a pair of coil springs and a pair of radius arm drop brackets as shown with this 4-inch lift. However, shocks as well as rear lift components also have to be purchased. Above 4 inches, something must also be done with the steering and the track bar. (Courtesy Superlift)

Track Bar

While both early and late-model Fords utilize a track bar, track bar angle is addressed in different ways. The older vehicles can benefit from an adjustable track bar, which is available from select suspension manufacturers. An adjustable bar allows the installer to properly center the axle under the vehicle, and as an added bonus they are rebuildable (unlike the factory bars). The most common track bar correction method on the late-model trucks is a relocation bracket at the axle or frame. Like all vehicles equipped with a track bar, it is important that the operating of the drag link and track bar match or poor handling traits will result.

What to Look for in a Lift System

First and foremost, verify that the kit includes everything needed. This consists of coil springs, shocks, track bar and steering correction (above 4 inches of lift), and caster correction. The overriding factor on the early trucks is not the base system components, but what else might be needed to end up with a safe vehicle once the lift installation is complete. This includes radius arm bushings, track bar bushings, steering linkage, wheel bearings, and ball joints. These components can easily

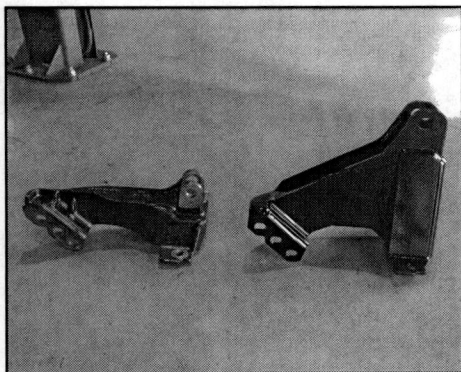

Late-model track bar correction usually happens via a drop bracket at the frame. The bracket on the right is a Superlift piece designed to be used with an 8-inch lift system and is much longer than the stock bracket on the left. Most Ford brackets are bolted in place, so there's no cutting or welding required.

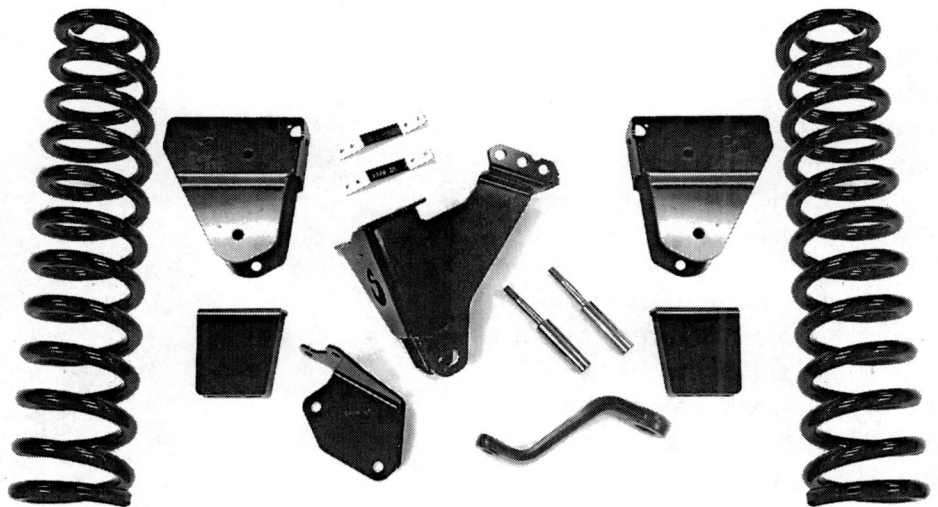

Late-model lift systems are more complete by necessity. This 6-inch Superlift kit for a 2005 Super Duty consists of new coil springs, radius arm drop brackets, a track bar bracket, a pitman arm, sway-bar links, a steering stabilizer bracket, and brake line drop brackets. Not shown are shocks and rear lift. (Courtesy Superlift)

What Fits, What Hits — Ford

Model	Year	Additional Modification	Tire Size (inches)								
			31	32	33	34	35	36	37	38	40
Bronco	1966–'77	None	3½	5½							
		Cut-Out Flare			3½		5½				
	1978–'79	None				1½	4	6½			
		Fender Trim					1½	4		6½	
F-100/150	1966–'79	None				1½	4	6½			9
		Fender Trim					1½	4		6½	
F-250	1977–'79	None					4				
		Fender Trim						4			
	2005-'07	None				2	4		6	8	
F-350	2005-'06	None				2	4		6	8	

double the overall price of the project, but a lift is only as good as the vehicle it is bolted to, so it's critical that the vehicle is in good working order. Also keep in mind not all of them had sway bars from the factory, so if your vehicle has one, plan on getting the necessary components that allow it to work properly.

Beyond the basics, things like progressive-rate coil springs are indicators of a quality lift system. Replacement radius arms offer improvements in both the looks and performance departments, but the factory components are more than adequate for a street vehicle provided caster is addressed via drop brackets or degreed C-bushings.

Many of the early Fords accommodate dual front shocks from the factory, but most kits are priced with only single shocks, so plan on factoring in an extra pair of shocks in your budget. Since both the early and late trucks are heavy, it's an excellent idea to ensure the kit includes high-quality shocks valved specifically for the application, and in many cases they benefit from a gas-pressurized shock.

Jeep Vehicles

Coil springs were first seen in 1984 with the introduction of the downsized Jeep Cherokee (XJ). At the time, no one anticipated just how big a part coil springs would play in future Jeep suspensions. The front suspension on the Cherokee is virtually the same as the later Wrangler TJ, so much so that some aftermarket products interchange between the two. With the introduction of the new Wrangler JK, Jeep went one step

Long a staple in the off-road community, Jeep took a bold step by going away from tried-and-true leaf springs to better riding and better-performing coil springs with the introduction of the 1997 Wrangler TJ. Viewed with skepticism at first, the TJ and now the new JK are often regarded as the most capable out-of-the-box 4x4s ever produced. This is in large part due to their coil spring suspension systems. (Courtesy Skyjacker)

The most economical way to gain a couple inches of lift on a TJ or JK is by placing urethane coil spring spacers atop the factory coil springs. These spacers are available from most lift manufacturers and allow additional clearance for an extra tire size or two. Still, even spacer lifts require new shocks in most cases. (Courtesy Superlift)

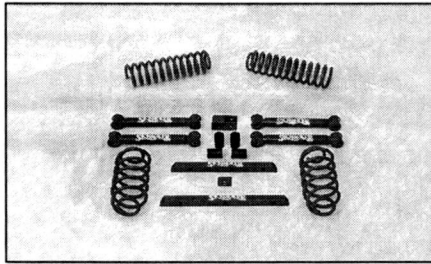

There's a threshold with TJ suspension systems that is crossed once you reach 4-inch territory. Basic 4-inch systems include replacement front and rear springs, a pair or a set of four lower control arms (to address caster, driveshaft, and wheelbase change), transfer case drop brackets, a track bar bracket, and compression stop extensions. These items are the minimum required to lift a TJ properly, but options and upgrades abound.

TJs, XJs, and JKs all use cam bolts to attach the lower control arms to the front axle. These cams allow caster to be restored after a lift is installed. For some inexplicable reason not all Jeeps come factory-equipped with cams, so they are often supplied with mild lift systems.

further and refined many of the shortcomings of the earlier suspension while keeping the rugged (and relatively simple) characteristics intact.

Lift System Basics

Due to the popularity of the Jeep vehicles using coil springs, the sky is pretty much the limit when it comes to suspension upgrades. Whether it's a boost of just a couple inches with some spacers or a high-performance system that replaces the factory suspension links with longer ones to substantially increase suspension travel, it's all readily available and the only limitation is your pocketbook.

For XJs and TJs, importance should be placed on caster correction. These vehicles use fairly short trailing links to locate the front axle,

so caster is lost quickly as lift increases. Generally speaking, mild 2-inch systems can get away with just cam bolts for the lower link arms. At 4 inches of lift, replacement lower links for the front should be considered the minimum acceptable. Most systems should also include replacement shocks and some form of driveline correction. TJs, even at 2 inches of lift, can develop driveline vibrations due to their short wheelbase and correspondingly short rear driveshaft. As with other solid axle designs, actual lift comes via either spacers positioned on top of the factory coil springs or replacement springs. The advantage of using spacers is that you can be sure factory ride quality is retained, but spacer lifts should be limited to a maximum of 2 inches. Replacement springs are a better option for those that intend to spend more than an occasional foray off-road. Compression stops should be addressed as well as the sway bar; the front sway bar plays an

important role with street handling and the early factory sway bar links are known to loosen up and cause irritating noises on higher mileage vehicles. Just like the previous Wranglers, there are conflicting views towards steering correction up to 4 inches of lift. Some experts say a drop pitman arm is mandatory at 4 inches of lift, while others claim a replacement pitman arm contributes to bumpsteer. Track bar correction is also important (more on this later).

Not a lot is yet known about the long-terms affects of a lift system on the new Wrangler JK, but the good news is that the design is basically the same as the previous generation except for the link arms that locate the axle, which are much longer than the TJ and, therefore, are more accommodating to mild lift systems. What is known is that just like the previous generation, caster is an important consideration. As a result, most basic systems include cam bolts for the front lower links, while others

Lift systems for JKs are not much different than the previous generation TJ and are often simpler thanks to the longer trailing arms used on the newer platform. As a result, 4-inch lift systems typically can get away with caster cams or replacement lower link arms for the front. Coil springs, track bar correction front and rear, sway-bar links, and compression stop extensions are standard fare. Amazingly, just 4 inches of lift clear up to 37-inch-tall tires thanks in part to the JKs extraordinarily large fenderwells. (Courtesy Superlift)

Four inches of lift is beyond the range of cam bolts to adequately restore caster. Therefore, most 4-inch systems provide replacement lower front trailing links that are slightly longer than stock. These links rotate the front axle slightly to gain the needed caster correction. Virtually all base-systems are simply a fixed link like this one and provide no additional suspension travel over the factory components.

include replacement lower arms to "roll" the axle and restore caster.

For the first time ever, the Wrangler is now offered in two- and four-door versions. Of course this comes with a large variance in curb weights between the two models, so look for companies that offer two- and four-door specific coil springs. Just like the TJ, replacement coils should be strongly considered by those looking to take their JK off-road.

Link Arms and Long Arms

Arguably the components most focused upon for upgrades are the trailing links that locate the axles. Factory XJ and TJ trailing arms are constructed of stamped steel with a rubber bushing at both ends. Simple and effective, they even have some "give" when flexing the suspension. However, they have been known to fail when used in extreme conditions and generally do not have the type of construction that inspires confidence. Most lift systems include a pair of front lower trailing links standard with a 4-inch system, and these links are slightly longer than stock to restore caster. Virtually all standard lift systems use a "fixed" link, that is, a tubular link with an eye welded at both ends and urethane bushings. Virtually imperceptible on the street, these links do their job quietly (provided the bushings are greased regularly) and without much fanfare, making them a perfectly acceptable choice for someone who does not intend to venture much beyond a graded dirt road.

The fun starts when thinking about link arms in terms of off-road use. As the Jeep's suspension flexes, a twisting force is introduced to the lower arms (as well as the uppers but to a lesser extent). The flexible fac-

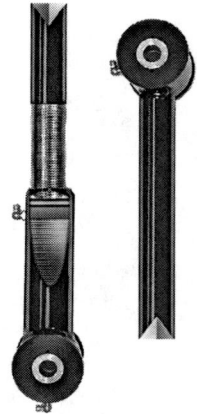

Superlift and Teraflex both solve the "TJ bind" with heavy-duty tubular links that have a threaded pivot in the middle. This design allows the link to twist as the suspension articulates and can also be used to fine-tune caster and driveline angle. Superlift's RockRunner links are available as a complete kit, or a standard system can be upgraded with them later on.

tory arm with rubber bushings have a certain amount of give that allows the suspension to flex, but tubular steel with urethane has very little give and quickly "shuts down" the suspension when one side is flexed much from ride height. Suspension designers figured out early on that creating a link capable of absorbing that twisting force was key to unlocking substantial amounts of suspension flex.

There have been many different approaches to building an arm that provides additional suspension travel, with each one having its benefits and drawbacks. Some are even unique enough to warrant patents. Some of the first replacement arms to come out utilized heim joints at one or both ends. A proven performer in various forms of motorsports, heim joints are durable, provide an impressive amount of adjustment, and provide all the flex necessary to be a good off-road performer. However, some companies use higher quality

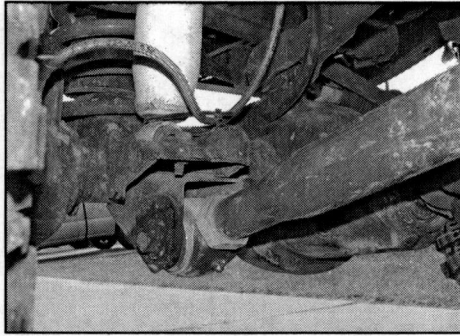

Replacement links that use heim joints like this well-used example are a popular choice among both the bolt-on and custom suspension crowd. This particular Skyjacker heim is massive and has a load capacity about triple what it is likely to experience in its lifetime. Skyjacker heims are also greasable to promote quiet operation and long life.

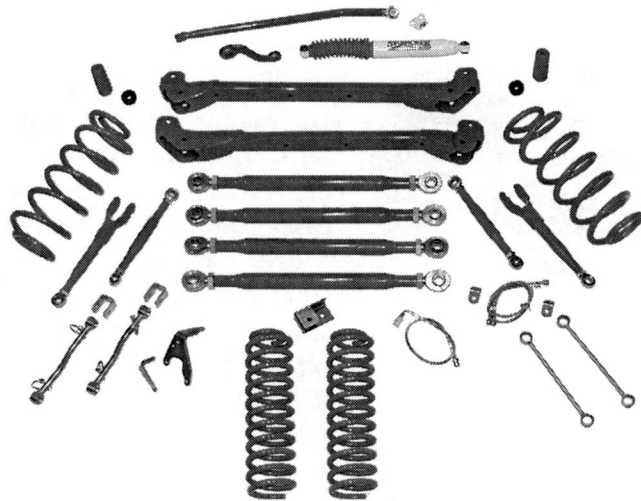

When it comes to selection that can accommodate any performance needs and budget, Skyjacker is hard to beat. They have systems from 2 to 8 inches of lift in three different levels. Its Value Flex systems appeal to the budget-conscious crowd, while its Double Flex system like the one shown here caters more towards the hardcore crowd. Skyjacker even has a coil-over conversion for those that want over-the-top performance. (Courtesy Skyjacker)

heims than others, and heims are known to become noisy over time, especially in dirt-prone environments. Another method is to use a uniball-type joint, usually with hard urethane inserts for noise absorption. These are typically adjustable (they can be tightened as they wear) and rebuildable. They have become quite popular among fabricators and have a proven record in race environments. However, noise can still be an issue when used hard and without regular maintenance. Still another design incorporates a threaded pivot in the center of the link. This style offers the same adjustability as heims as well as quiet operation as long as the links are serviced regularly. Still, the threads in the pivot have been known to gall and lock up if not greased regularly. If the pivot is damaged the entire link must be replaced, and some of these links usually use urethane bushings that tend to squeak if not serviced regularly as well.

On the more extreme end of the spectrum, long-arm systems have become the hardcore off-road standard for TJs. With a long-arm system, the factory upper and lower arms are replaced with much longer arms that typically connect to a new belly pan under the center of the Jeep. Long-arm systems offer incredible amounts of flex and are usually built to withstand severe off-road use, making them the ultimate choice for the hardcore enthusiast. However, most long-arm systems have been designed more around off-road use and sacrifice some on-road driveability characteristics. Long-arms are cool and perform well, but they are also very complex to install and easily triple what a conventional lift system costs. We'll take a closer look at long-arm kits in Chapter 9.

As mentioned earlier, the new JK utilizes much longer trailing links than the older JK, making them a "semi-long-arm" right from the factory. Though it's still early in the JK's

life, suspension manufacturers are already hard at work developing better link-arms for off-road use, and it's a safe bet that much of what has been learned on the TJ will be applied to this new platform.

Driveline Angle

Jeeps are by nature short and nimble, which is one of the reasons they are such popular off-road vehicles. However, a short wheelbase means short driveshafts, and just as the operating angles of short trailing arms increases dramatically, so can the driveshafts. Driveline angle and vibration have plagued all short-wheelbase Jeeps since they were invented, and the TJ is no different. The availability of many different swappable engine, transmission, and transfer case combinations only aggravates the issue. With as little as 2 inches of lift, it may be necessary to address rear driveline angle. Some companies include cam bolts that replace the bolts for the upper link

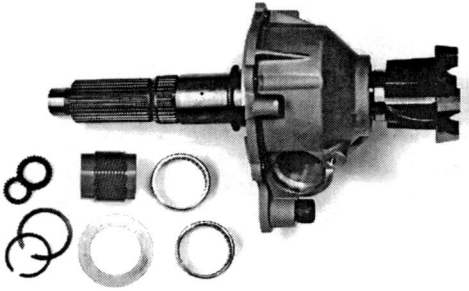

An SYE kit is an upgrade that should be strongly considered on any YJ or TJ that gets used off-road. Several different types are on the market, but the better ones utilize a heavy-duty mainshaft that replaces the failure-prone factory shaft. Then again, the less expensive versions that retain the factory mainshaft do not require disassembling the transfer case.

arms at the axle. These cams allow the installer to raise the pinion in order to "dial-in" the proper angle. At 4 inches of lift, however, something must be done. The easiest and most common correction involves spacing down the transmission crossmember, which in turn lowers the transfer case. Though effective, this does reduce ground clearance and hinder off-road prowess.

A better, though more expensive, option is to install a readily available Slip-Yoke Eliminator (SYE) kit in the transfer case. A relatively simple task for the mechanically inclined (installation details are included in Chapter 8), an SYE kit involves replacing the factory driveshaft with a slip-spline at the transfer case with a more conventional shaft that utilizes a constant-velocity (CV) joint and a conventional slip-spline built into the driveshaft. The net effect is increased driveshaft length, the ability to utilize a conventional CV that can handle more driveline angle,

and the ability to continue operating the vehicle in the event of rear driveshaft failure (if the factory driveshaft fails, continuing to operate the vehicle will result in the loss of transfer case fluid). SYE (sometimes also called "fixed-yoke") kits are available from any aftermarket supplier that specializes in Jeeps and from several lift companies. In most cases, an SYE kit eliminates the need for a transfer case drop at 4 inches of lift for TJs, TJDs, and XJs. Some of the better kits even eliminate the need for a drop at 6 inches of lift.

One interesting note on the JK is that driveline angle is less of an issue due to the vehicle's longer wheelbase, even in the 2-door versions. As a result, lowering the transfer case is not necessary at 4 inches of lift *so far*. Unfortunately, the JK also utilizes "tulip" style joints at the transfer case for both the front and rear driveshafts. Though new to the Wrangler, these joints are known for being sensitive to even minor changes in operating angle, with accelerated wear being evident on other applications that use them (such as the Jeep Liberty and late-model Ford Ranger). At the time of this writing, these vehicles are simply too new to know the long-term effects of uncorrected driveline angle on the JK with lifts up to 4 inches. In the short term no vibration is evident on low-mileage vehicles, and driveline companies have already started releasing conversions that allow the use of "conventional" driveshafts.

Track Bar

Though it has been covered before, it bears repeating that the goal of any track bar correction is to properly center the front axle under the vehicle and match the operating

Apparently not learning the lesson from the poor TJ rear driveshaft design, the new JK utilizes what is commonly called "tulip" style joints at both ends of both driveshafts. Though it is too early to tell how these will hold up with a lift system, other applications that use this type of driveshaft joint are known to show premature wear when operated at greater angles than factory. Fortunately, driveshaft upgrades are already available from the aftermarket. It just remains to be seen if costly driveshafts will be needed long-term on a lifted JK.

angle of the drag link. Though the previous generation YJ had track bars front and rear, the leaf springs did the majority of the axle-locating duties side-to-side. On TJs and XJs, the front track bar is a critical component. With basic systems, the side-to-side correction is done by moving its attachment point at the axle via either a relocation bracket or simply by re-drilling the factory mount at the axle. This allows the factory bar to be retained; however, a replacement adjustable track bar should be strongly considered with these vehicles. The factory bar utilizes a very wear-prone ball-type joint very similar to a steering joint. These are notorious for premature wear and are subsequently the source of steering slop and even death wobble. A replacement track bar, available

Track-bar operating angle is an important aspect of ride and handling on coil-spring Jeeps. If you compare the angle of the drag link (the steering link attached to the pitman arm) and the track bar (the unpainted piece in the background), you will notice that they are parallel with one another. Track-bar phasing is tricky and can take some trial and error to set up properly as evidenced by this prototype piece on a Superlift system.

Because steering is not a consideration, most kits for TJs and JKs supply a relocation bracket for the rear track bar. New adjustable bars are also available for these applications.

This Superlift JK system utilizes a replacement adjustable track bar to handle properly centering the front axle, but its operating angle is unchanged to match the angle of the drag link attached to a factory pitman arm. Changing one or the other but not both has been shown to create bumpsteer on earlier platforms.

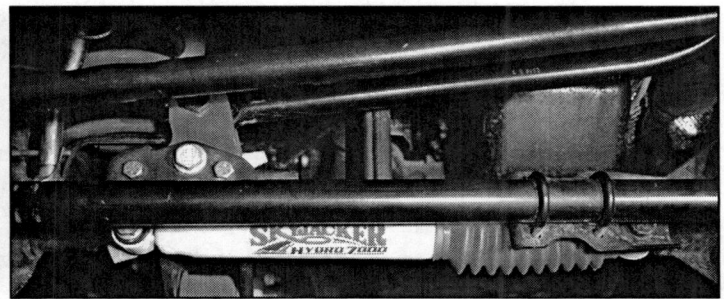

Skyjacker's approach to track bar correction on the JK is a simple bracket that raises the attachment point at the axle. Still, it appears the drag link runs at the same angle, so bumpsteer concerns may be unfounded. More likely, either approach is acceptable as long as something is done at 4 inches of lift on the JK. (Courtesy Skyjacker)

separately with many 4-inch systems, allows the factory mounting points to be retained. This design enables precise centering of the front axle and eliminates the factory bar (which is not rebuildable) with one that is serviceable. With most 6-inch systems, replacement track bars are needed front and rear. When used in conjunction with a drop pitman arm, most handling pitfalls can be avoided.

As for the JK, there are already different camps forming in regards to track bar correction and method. While most experts agree that a pitman arm is not needed at 4 inches of lift and below, at issue is modifying one or both attachment points for the factory track bar, or supplying a replacement track bar that is longer but utilizes the factory mounts at the axle and frame. While each method has its arguable advantages, conventional wisdom holds that moving the track bar via a relocation bracket without addressing drag link angle can lead to issues with bumpsteer. At the same time, this may also be one more case where practical experience trumps traditional thinking. Only time will tell. Regardless, the good news is that the JK does not use the problematic track bar design found on the XJ and TJ.

Steering

It was hinted at in the track bar section, but steering correction is somewhat controversial on the XJ

Opinions differ on the use of a drop pitman arm on TJs with 4 inches of lift; some kits include one, while others don't. However, 6-inch systems require the use of a drop arm like the one shown here.

Superlift's standard 4-inch lift system is typical of most base 4-inch kits for TJs. (Courtesy Superlift)

and TJ. Some lift manufacturers strongly recommend the use of a pitman arm at 4 inches of lift, while others consider it optional. There are well-respected off-road experts in both camps as well. The solution most likely has to do with track bar *and* steering correction. If one is addressed, so must be the other. The best advice here is to follow whatever recommendations the suspension manufacturer has in regards to steering correction, as there will be some recourse if some of the common complaints develop after lifting the vehicle. At 6 inches of lift and above, everyone agrees that a pitman arm is needed. On the JK, it is safe to assume that 4 inches of lift requires no steering correction, while 6

inches of lift does. However, with the ability to clear up to 37-inch-tall tires with a 4-inch system, some question the need to go any higher.

What to Look for in a Lift System

Much of the information on what to look for in a lift system for coil-spring Jeeps can be gleaned by reading the other contents of this section. Suffice it to say that studying the different problem areas covered and then making sure the lift in question has the necessary components to address these areas is the best advice. Avoid bare-bones kits that don't include things like lower link arms for caster and driveline correction. Better systems include both front and rear lower link arms for TJs

to also correct wheelbase and rear driveline angle. As with other platforms, taller lifts require more items to work properly, and this can add expense. Making sure everything needed is included (and not listed as optional) is the best defense against surprises.

Dodge Trucks

In quite possibly the biggest role reversal in automotive history, Dodge went from the most dated, stodgy, and outdated truck manufacturer to the boldest and most aggressive virtually overnight. The year was 1994, and aside from radical sheet metal changes (the design influences of which continue to this day), the

What Fits, What Hits — Jeep

Model	Year	Additional Modification	Tire Size (Inches)								
			31	32	33	34	35	36	37	38	40
Cherokee XJ	1984–'01	None	3	4							
Wrangler TJ, TJD	1997–'06	None		2	4		6				
		Body Lift 2"			2		4	6			
Wrangler JK	2007	None			2		3		4		

new Dodge sported a whole new coil spring suspension. Fast forward to present day and the ¾-ton and 1-ton models still utilize the same basic suspension design that was introduced over a decade ago.

Suspension Basics

Dodge utilized the same suspension on its ½-, ¾-, and 1-ton trucks from 1994 to 2001. In 2002, Dodge succumbed to peer pressure and the ½-tons were changed to independent front suspension in order to achieve the same ride and handling characteristics as its competitors. The basic configuration is a solid front axle located front-to-rear via upper and lower trailing arms, and laterally by a track bar. Steering is handled with a conventional steering box and a typical drag link/tie-rod arrangement. The '94–2001 models shared the same components, with only heavier springs and larger front axles distinguishing the ¾- and 1-ton models from their ½-ton counterparts. There was a major redesign in 2003 that included minor changes to the suspension (most notably addressing the problematic track bar on the previous generation), but by and large the suspension remains the same. The few changes to the original design are a testament to its durability, and it performs well on-road and off. As an added bonus, lift systems can provide enhancements in suspension travel as well as articulation, in addition to more room for larger tires.

Lift System Basics

Like other coil spring lift systems, actual lift comes from the installation of taller coil springs or coil spring spacers, but just like all the others, it's not quite that simple. Dodge trucks are particularly suscep-

Whether it's a play toy or a heavy hauler for play toys, Dodge trucks fit the bill for many off-road enthusiasts. (Courtesy Skyjacker)

tible to even minor caster changes, so once the lift is above the 3-inch range, lift designers must maintain a delicate balance between caster and front driveline angle.

A basic lift system for these vehicles consists of replacement coil springs, a track bar bracket, a drop pitman arm, bumpstops, sway bar links, shocks, and either a pair or a complete set of four replacement trailing arms. Fairly basic as lift systems go, the design's relative simplicity makes lifting the solid-axle Dodge more affordable than many other suspension designs.

Of course, the complexity of the system depends on lift height. Two-inch systems can include as little as a pair of coils or coil spring spacers and shocks. At 3 inches, a track bar bracket

The 2003 and newer Dodge front suspension is remarkably similar to the previous generation, with only minor tweaks to address some shortcomings of the earlier design. Still, Dodge trucks are very receptive to lift systems up to 6 inches.

is generally added, with spacers still being an option among a few manufacturers. A 4-inch system is where the replacement trailing arms make

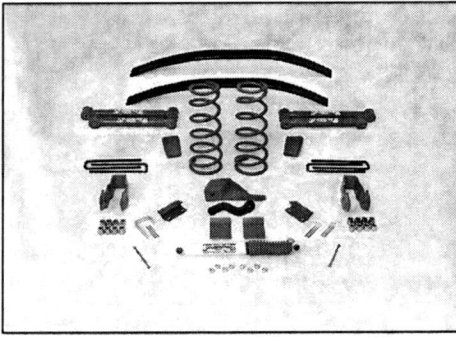

This 4-inch lift for '94–2002 trucks is typical of the standard lift systems out there. Replacement coil springs provide lift, while new link arms restore caster on the front end. A track bar bracket spaces down the factory bar at the frame, and a pitman arm handles steering duties. Bump stop brackets keep damaging over-compression in check, as do replacement urethane bump stops. Add-a-leafs for the rear round out this basic but very complete kit. Of course, Skyjacker offers a multitude of upgrades like many other manufacturers. (Courtesy Skyjacker)

A simple way to gain a couple inches and level the stance of the truck (Dodge trucks have a more substantial rake than most), coil spring spacers are a viable option to step up a tire size or two. Because of the minor lift involved, nothing else is needed and factory ride quality is maintained. (Courtesy Skyjacker)

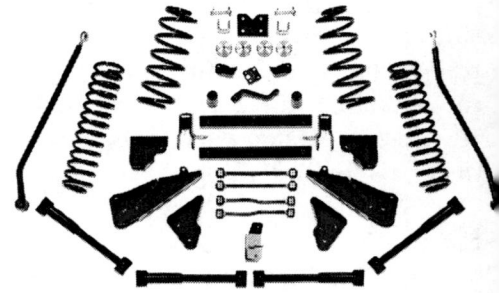

Comparing the 2-inch spacers with this Superlift 6-inch system reveals just how dramatically different 4 inches of lift can be in terms of complexity. Still, all of this stuff is needed in order to reach loftier heights safely. (Courtesy Superlift)

their appearance, along with bumpstops, a pitman arm, and sway bar links. There are a few systems on the market that offer lifts higher than 6 inches, and most are well designed. However, be aware that front driveline issues sometimes arise at these taller lift heights that can require costly replacement driveshafts and/or a custom axle if the vehicle is going to be used hard off-road.

Link Arm Options

Just like the Jeep TJ (which shares a very similar suspension design), lift manufacturers often tweak the trailing arms in order to unlock hidden performance gains. Like the TJ, a twisting force is placed on the trailing arms as the suspen-

sion is loaded on one side. Bushings absorb this twist to a certain extent, but the trailing links quickly start to bind, which eventually prevents the suspension from articulating any further. By simply eliminating this bind, suspension articulation can increase dramatically.

Most basic lift systems do not address this issue, as it does not impact on-road driveability. Typically, replacement tubular trailing arms are supplied with urethane bushings that can actually improve cornering ability because urethane bushings are stiffer than the rubber bushings they replace. For the upgrades systems, like the TJ, suspension designers go about eliminating trailing link bind a variety of different ways. Some use replacement links with heim joints at one or both ends, while others utilize an articulated link. The benefits and drawbacks of these different designs are briefly discussed in the TJ section elsewhere in this chapter. The important thing to remember with these

systems is not to lose sight of the intended use of the vehicle. Look, considerations aside, if the truck is going to see only street duty, a basic system with fixed trailing arms is a perfectly acceptable choice because the vehicle will experience little if any benefit using the "upgraded" arms. The one exception is that most arms that offer more articulation also can usually be adjusted to dial in caster and driveline angle.

For those that want to take things one step further, select companies also offer "long-arm" systems for most models of coil-sprung Dodges. A long-arm system offers obvious suspension travel gains, but also allows the trailing arms to remain more or less parallel with the ground (which improves ride quality) and reduces the amount of caster change as the suspension cycles. Of course, these systems are a premium over the conventional systems, but the off-road performance gains are worth the extra expense provided they are actually utilized. Long-arm systems can also accommodate additional lift for the show truck crowd.

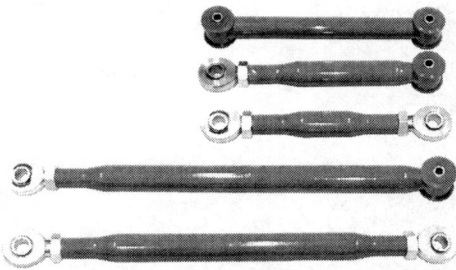

Just like its Jeep product line, Skyjacker offers three different levels of replacement link arms to suit various budget and performance needs. The base systems consist of fixed links with urethane bushings at both ends. Single Flex systems have a heim at one end and a bushing at the other, while top-of-the-line systems have heims at both ends to soak up twist. At the high end of the spectrum, long-arm systems provide the ultimate in off-road performance, but this comes at a premium. A few suspension manufacturers offer one or two levels of upgrade arms like these Skyjacker offerings. (Courtesy Skyjacker)

Taking a different approach but still applying its Jeep technology, Superlift offers optional RockRunner arms that utilize a pivot in the center to soak up flex (the pivot is near the grease fittings visible on the arm). Though subtle, these arms offer quite a bit more suspension articulation. Also visible are some of Superlift's other options, including a multi-shock kit.

Above 6 inches of lift, a long-arm system is really the only option. Available through Skyjacker, Fabtech, and a few others, Dodge long-arm systems are the extreme end of the spectrum. Keep in mind that front driveline angle becomes a concern, and Skyjacker addresses this by re-clocking the transfer case via an indexing ring. Hybrid driveshafts are also a viable option for those sensitive to ground clearance reduction.

Track Bar Concerns

It has been covered elsewhere in this chapter, but it bears repeating here that ensuring the track-bar operating angle matches that of the drag link is critical. This is generally accomplished via a bracket that lowers the track bar attachment point on the frame and a drop pitman arm. The main thing to keep in mind is that although it may look easy to "upgrade" a lift from, say 4 to 6 inches, usually the track bar also needs to be changed to match the new lift height. If left unattended, bad handling traits will inevitably result.

One special note must be mentioned regarding the track bars used on the '94–2002 Dodge trucks: they're well known to wear quickly, and the stress of larger tires only aggravates them further. Similar to the TJ, this era of Dodge truck utilizes a ball-joint type of end at the frame that is often the source of strange handling characteristics, death wobble, and sloppy steering. A worn track bar is less noticeable when the vehicle is stock, so it is not uncommon to hear about a vehicle that handled fine at stock height and then suddenly developed all kinds of handling problems when the lift was installed simply because the track bar was out of spec. Unfortunately, the track bar is not serviceable and must be entirely replaced. The moral of the story is to check the track bar carefully, and replacing it on a high-mileage vehicle when installing a lift is safe insurance.

What Fits, What Hits – Dodge

Model	Year	Additional Modification	31	32	33	34	35	36	37	38	40
½-Ton	1994–'01	None			2½		5				
		Fender Trim				2½		5			
¾, 1 Ton	1994–'02	None				3	5				
	2003–'06	None			2		4		6		

Ram It With a 6-inch System

Dodge trucks were nearly extinct by 1992; aging technology and stodgy styling had relegated Dodge trucks near the bottom of the popularity ladder. Recognizing it was "go" time, Dodge engineers went waaaaaay outside the styling box and released their revolutionary-at-the-time 1994 Ram... and they've never looked back. Long since copied by more than one manufacturer, the big-rig styling cues of the current Ram, along with some significant equipment offerings, have kept Dodge firmly near the top of the big-truck market. Dodge truck owners are a fiercely loyal and enthusiastic bunch, so it's probably not a coincidence that Ram lift systems are the top sellers for several aftermarket suspension manufacturers.

Recognizing that it is unwise to mess with a good thing, Dodge trucks have been subject to only minor suspension and styling changes since the '94 model, meaning that lifting them is the same with some minor exceptions. Nearly all major suspension companies offer something for

Big and bad, this Dodge is ready for the trails thanks to 6 more inches of clearance and bigger tires to match.

the Ram, with lift heights ranging from 2-inch leveling systems all the way up to massive 7-inch kits for certain models. The suspension itself is mercifully simple, which means it's also easy to modify. Someone can install virtually all systems offered with average-or-above mechanical experience and a reasonably stocked toolbox.

To see what it takes to lift a late-model Ram, we traveled to Robby Gordon Off-Road and followed along as its expert technicians gave the full Superlift treatment to a 2003 Ram 2500. The install took less than a day and required no welding; this is strictly bolt-on stuff and there's not even much drilling to be done. Furthermore, Superlift's 4-inch lift is virtually identical but with shorter coils. By the end of the day, the truck was rolling on 37-inch tires and had a whole new stance to match its style.

1. Superlift systems are some of the most complete to be found with gas- or diesel-specific coil springs, replacement upper and lower control arms, a track bar bracket, a pitman arm, urethane bump stops, sway-bar links, blocks for the rear, and your choice of three different shock options. The 4-inch system is virtually identical. (Courtesy Superlift)

2. Taking proper safety precautions, remove the necessary stock components. This includes disconnecting the track bar, drag link, sway-bar links, shocks, and then lowering the axle enough to remove the factory shocks and springs. Superlift's detailed instructions cover the entire procedure from start to finish.

Ram It With a 6-inch System *continued*

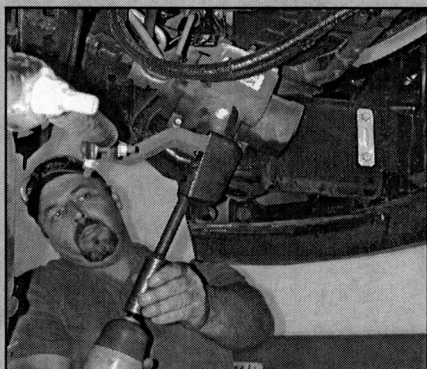

3. The kit includes a new dropped pitman arm, and a special puller tool is required to remove the old one. These can be rented from local parts stores for little-to-no money, but be aware that Dodge arms in particular are stubborn about coming off. Be sure you take the necessary precautions to avoid injury should anything slip or break during the pulling process.

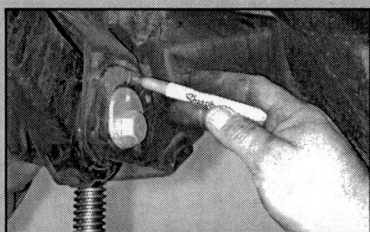

4. When comparing the new arm (already installed and torqued) with the old one, you can see that the new arm places the drag link attachment point about 3 inches lower than the original, relieving the operating angle of the drag link to work with the lift system.

5. Like most systems from other manufacturers, the Superlift standard kits include replacement heavy-duty tubular trailing arms that provide the necessary caster correction. The lower arms have cam bolts at the axle for alignment adjustment purposes. It's a good idea to mark the cams' orientation before removing them for reference during assembly. Installing the arms should be done one side at a time so that the bulky front axle is always at least partially attached to the truck.

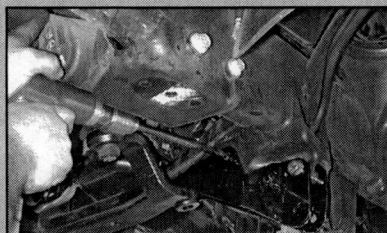

6. One option that off-roaders should seriously consider is Superlift's RockRunner arms. These trailing arms replace the factory pieces just like the fixed arms in the standard kits, but the RockRunner arms have a pivot built into them that allows the link to twist as the suspension flexes. This twisting ability reduces suspension bind and unlocks substantial amounts of suspension articulation. Furthermore, these links allow caster and driveline angle tuning.

7. Just like the drag link, the factory track bar must be lowered to compensate for the lift. If the angles of these two items don't match, serious handling issues will result. Superlift's track bar bracket uses two factory holes and requires drilling a third one. There's no guesswork or measuring here, just line it up and bolt it on, which is as things should be.

8. It's important to control compression travel with a coil spring suspension because allowing the springs to over-compress and bind fatigues them and causes premature sagging. Compression stops also help the suspension have a more progressive rate. Superlift's urethane 3-stage bump stops and mounting brackets keep the front axle from compressing too far upward, which saves the springs and keeps the tires out of the fenders.

Ram It With a 6-inch System *continued*

9. Another beneficial option for all coil suspension designs is a multi-shock system. Superlift's design allows up to two additional shocks to be added per wheel. First, lower brackets are attached to the link arm mounts via a couple of drilled holes.

10. The tubular shock hoop looks pretty cool and installs by drilling only one hole; all the other holes already exist on the frame. A number of shock mounting options exist with this design: A dual setup attaches to the hoop with no shock in the factory location inside the coil, one on the hoop and one in the coil, or a triple system like what is shown here. The owner of this truck chose to use Superlift SS gas-charged shocks rather than the standard Superide versions. Though they look pretty cool, duals are more than adequate for anything other than racing. As you can see, it's necessary to install the shock and the coil together if the stock location will be used.

11. Just like the shocks, a number of different steering stabilizer options are available. The basic version utilizes the factory locations, or you can step it up with a dual stabilizer system. If the dual route is chosen, it's necessary to cut off the factory stabilizer mount on the axle.

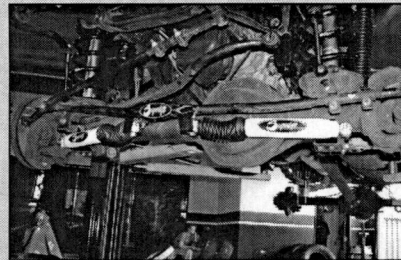

12. The finished product once again looks better than the stock stuff, though this style does hang below the tie rod where the stabilizers can get whacked on obstacles. With the big tires this kit enables the truck to clear—dual stabilizers really help keep those big meats under control.

13. Coil suspensions are well known for needing a sway bar and the Ram is no exception. The kit includes longer sway-bar links, which is preferable to lowering the bar from the frame via bracketry. The links are also greasable and should be serviced with every oil change.

14. Something that is often overlooked by lift manufacturers, the emergency brake cables often need to be relocated on taller lift systems. Fortunately Superlift has the situation covered with an e-brake cable bracket that repositions the tensioner to compensate for the 6-inch lift. It simply bolts to one of the body mounts as shown.

Ram It With a 6-inch System *continued*

15. The stock rear springs are well engineered and extremely long, making them flexy right out of the box and very expensive to replace with lifted units. As such, tapered blocks and extended U-bolts bring the rear up to match the front. These blocks are also twice as heavy as normal lift blocks, so they should serve well even behind diesel torque.

16. Yet another item not to be overlooked, the rear bump stops play an infrequent but important part in the overall rear suspension. Superlift supplied a simple bracket that spaces them down enough to engage the axle before sheet metal begins wrinkling when it meets rubber.

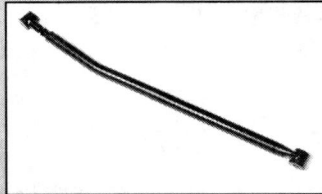

17. How about some rear dual shocks to match the front? The Dodge makes it easy to add an extra shock at each rear wheel, which helps with the rough stuff, but can impact ride quality unless the shocks are valved to work in a dual application. They are in this case, but dual shocks are easy to add yourself; all it takes is some longer bolts and a few washers.

18. Cummins engines produce tremendous amounts of torque, and torque plus lift blocks often means axle wrap. To head this potential problem off at the pass, the owner of the truck opted for Superlift's optional Torque Max traction bars. These bars are designed to eliminate axle wrap while allowing the rear to flex normally thanks to the use of a threaded pivot and careful attention to proper geometry. They even have a stainless insert, so they look cool, too.

The finished product provides ample room for 37-inch-tall tires and a stance to match the truck's aggressive styling. A quick off-road foray revealed newfound flex and a spring rate well suited to high-speed whoops. Not only that, the truck now garners looks wherever it goes.

Twin-Traction Beam (TTB)

This Chapter Includes:

'80–'96 Ford F-150
'80–'96 Ford Bronco
'80–'97 Ford F-250
'84–'90 Ford Bronco II
'90–'94 Ford Explorer
'83–'97 Ford Ranger

Twin-traction beam may be controversial, but it can be made to perform well in off-road environments. Best of all, TTB trucks are cheap.

In the history of four-wheel-drive vehicles there is no suspension design more controversial than Twin-Traction Beam (TTB). Often considered the redheaded stepchild of the off-road world, it is as much reviled by the slow-speed off-roaders as it is cherished by the racing crowd. TTB inspires hatred by many, and even the racers agree that there are significant issues with the design in stock form. It will forever awkwardly straddle the line between solid axles and IFS because it's technically the latter while having an odd combination of attributes and drawbacks among both suspension styles. Even so, the suspension design was innovative at the time and did manage to carve out a little respect in the off-road world.

As the application list indicates, Twin-Traction Beam was utilized extensively (and exclusively) by Ford in its two- and four-wheel drive light trucks (2WD versions were referred to as Twin-I Beam). That's right, the two-wheel- and four-wheel-drive suspensions in the eras listed above are nearly identical, with the only difference being steel beams taking the place of stamped steel differential halves. Though Ford probably originated the design of the front suspension, the front axle itself uses common parts manufactured by Dana.

The reasoning behind TTB is sound enough: combine the simplicity and reliability of a solid front axle with the undeniable ride quality advantage of independent front suspension. However, the execution is a bit bizarre and creates a whole new collection of quirks not found on other suspension designs. Still, when looking at TTB as objectively as possible, it does cater to both the on- and off-road worlds by having a

Whether it's the current mud-racing scene or Baja classes both old and new, TTB, or at least the concept, attracts racers to this day. This is a mud-racing truck pulling more than 700 horsepower on a horseshoe-shaped mud-racing track in Louisiana. TTB allows the truck to corner more efficiently and handle the high-speed bumps present in the racecourse better than its solid-axle competitors. (Courtesy Skyjacker)

Despite its challenges, TTB has managed to gain some respect in the world. Properly modified, it can serve as both daily driver and off-roader.

comfortable ride on the street and a flexy, capable, sure-footed suspension off the road. Though not without some inherent design flaws and strength reductions, TTB can be successfully modified to be an adequate performer with larger tires both on and off pavement.

How It Works

The only really funky part to TTB is the axlehousing itself. Instead of one solid housing, there are two axle halves that pivot on brackets mounted to a very stout engine crossmember. Further out, the halves are located by short stamped steel radius arms and suspended by coil springs just like the older Ford trucks. The internals of the pivoting axle assembly are surprisingly similar to their older solid axle counterparts, with the ½-ton axles sharing many components (including Dana 44 gears and carriers, ball joints, stub shafts, and axle U-joints). The main exception is the axleshafts themselves, the passenger side of which has an extra joint to accommodate the whole pivoting assembly. Basically, it is a solid axle chopped in half. As for the rest of the assembly, the radius arms use "bayonet" style bushings like the older Fords that are well-known performers, and many trucks accommodate dual front shocks. The steering, on the other hand, is a fairly undesirable arrangement with a long drag link running from the pitman arm to the passenger knuckle and a fairly short tie rod connecting the driver side knuckle to the drag link. This arrangement is more or less identical on the mid- and full-size trucks, with the one exception being select models of F-250s that are suspended by leaf springs.

So why go through all of this trouble? It is clear that TTB is much more complex than the solid-axle designs they replaced, and more stuff means more expense, right? Yes, but the real reason is all about ride quality. By incorporating a central pivot, a road irregularity encountered by one side of the suspension does not affect the opposite side. With a solid axle, the two front tires are rigidly con-

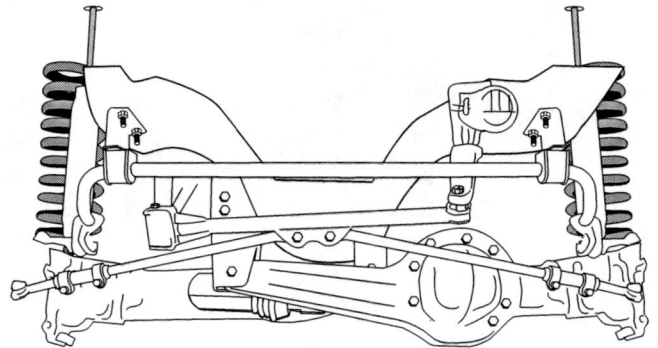

A typical TTB suspension using coil springs. Three-quarter-ton variants used leaf springs rather than coils, as did a select few 1-tons.

A typical standard TTB 4- or 6-inch system includes new front coil springs, axle pivot brackets, radius arm drop brackets, a pitman arm, sway bar drop brackets, and shocks all the way around. Though certainly adequate, numerous options and upgrades abound for better driveability and off-road performance. (Courtesy Superlift)

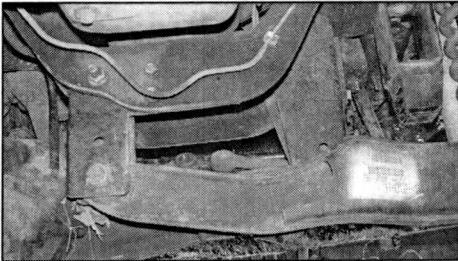

New axle pivot brackets are key to any TTB system. These brackets restore camber and correct the operating angles of the axleshafts all at the same time. Note that most manufacturers utilize the same brackets for 4- and 6-inch systems, and these brackets have two sets of mounting holes for the axle. For 4-inch systems, be sure to use the upper holes, while 6-inch systems use the lower holes as shown here.

nected, so hitting a pothole with one tire is going to be felt by the rest of the suspension (and, therefore, more easily transferred to the passenger compartment). With TTB, hitting a pothole with the passenger tire only impacts the passenger-side spring and shock instead of the whole sus-

pension, so the passenger side can deal with the impact while the driver side simply goes on about its business, oblivious to what is going on with the other side, and vice versa.

Okay, so if ride-quality is the reason, why not just make the leap to true independent suspension using control arms like the designs that came later? There is no firm answer to this question, but one plausible explanation is that solid axles have a faithful following in the off-road market, so perhaps Ford attempted to convert the non-believers by using something that appeared very similar to past equipment while taking a stab at improving ride quality. At the same time, it is not a coincidence that using largely existing axle parts saved substantial amounts of money for the company.

Lift System Basics

Because a TTB is more complex than a solid axle, so too is lifting the suspension safely. But even with the extra brackets, lifting a TTB is still less

expensive than raising the IFS that eventually replaced it. The overriding critical issues with TTB are controlling caster and camber, because both are substantially impacted when modifying this system for more ride height. More on this later…

Virtually all reputable lift companies address both caster and camber via drop brackets. A pair of brackets lowers the attachment points for the axle halves at the frame (for camber), while another pair of brackets lowers the attachment points for the radius arms at the frame (for caster). The rest of the components are standard fare, including replacement coil springs, shocks, a pitman arm (at 4 inches of lift and above), sway-bar drop brackets, and compression stop extensions. With just 2 inches of lift these drop brackets can generally be avoided, with alignment issues resolved by an alignment bushing that adjusts the position of the upper ball joint. At 4 inches of lift and above, drop brackets for key suspension components are required.

In addition to the basics, options are plentiful. These include multi-shock systems, extended radius arms, and even semi-custom race parts that can be adapted for everyday use. Beyond the alignment concerns, steering geometry must also be addressed. The factory linkage is not well suited to lifting due to its design, but there are fixes out there. More on this later as well…

Alignment Matters

Probably the most common complaint with TTB is its apparent inability to maintain proper alignment. Seemingly every time one turns around, some irregular tire wear becomes evident or the han-

Adjustable caster/camber shims are available from a few lift manufacturers that provide a greater degree of adjustment than those that a typical alignment shop has access to. These shims change the location of the upper ball joint in relation to the axle and operate in a similar principal to a cam bolt for a link arm. These cams offer up to 3¼ degrees of correction for caster, camber, or lesser degrees of both.

Since it might be difficult to understand exactly how the caster/camber shims work, here you can see one installed under the retaining nut for the upper ball joint. You might notice the slope in the angle of the nut in relation to the rest of the axle; that's because the shim in this case is making a strong adjustment for camber due to some heavy accessories on the front of the truck.

dling of the vehicle gradually heads south. The reason alignment can be so problematic with TTB is that several different components can impact alignment—some immedi-

ately, and some over time. The immediate fixes are built into any reputable lift system, but the wild card is ride height.

Camber

As mentioned earlier, both caster and camber are profoundly affected with TTB once a lift comes into play. While caster has been discussed extensively throughout this book, up to this point camber has barely been mentioned. Why does this critical alignment specification arise all of a sudden? The answer has to do with the pivoting axle halves. As a refresher, when viewed from the front of the vehicle, camber is the inward or outward tilt of the top of the tire in relation to the bottom. With the two solid axle designs the book has covered so far, camber is built right into the construction of the axlehousing. There is actually a small amount of camber adjustment possible in most front axles, but this is usually done at the factory during the construction of the vehicle and is never touched again for the life of the truck (which is why some may be surprised to learn that it is even possible to make camber adjustments to a solid axle). The pivoting axle halves change the whole ball game. The halves are basically levers with a fulcrum on one end and the tire at the other. If a taller spring is installed without moving the pivot at the frame, the lever is forced downward, which in turn forces the top of the tire outward in relation to the bottom of the tire. Conversely, installing a spring with less height (such as one that has sagged) forces the top of the tire inward. Camber impacts a variety of driveability traits (such as cornering) but also has a profound impact on tire wear. Driveability con-

Caster correction is all about the radius arm attachment point on the frame. With stock radius arms the attachment point is lowered from the factory location. Extended radius arms, like this one, both move the attachment point further back on the frame and incorporate caster correction into the design of the arms themselves. If you look closely, you can see one of the original mounting bolt-holes for the radius arm bracket between the two legs of the body mount in this photo.

cerns aside, off-road tires tend to be expensive and improper camber can ruin a set of tires in short order. This is why it is important to keep camber within acceptable limits.

The basic fix for camber is ensuring that the axle pivot brackets (for the axle halves) match the intended lift height. They *are* specific to the lift height, though many lift manufacturers produce pivot brackets with two different sets of holes in them for 4- and 6-inch systems. In most cases, a 4-inch lift should place the pivot bolts in the upper holes, while the 6-inch system places them in the lower holes. One other approach to addressing camber is utilizing cam bolts for the axle pivots, which allows an alignment technician to dial in camber and allows future adjustments as the springs settle. Other manufacturers use adjustable

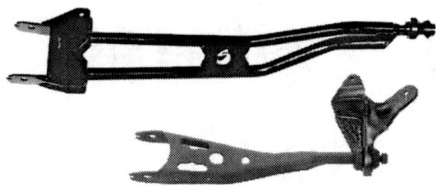

When comparing the factory radius arms with a drop bracket and aftermarket extended-length replacement pieces, it is apparent that quite a bit of ground clearance is gained with the aftermarket parts. However, the benefits don't stop there: better tire clearance, caster correction, suspension travel, and more can be had with this simple upgrade.

alignment cams that serve the same basic purpose and can be used for both camber and caster.

Caster

Equally as important as camber, caster is a big deal when dealing with TTB. Picture the radius arms as a lever with the fulcrum (at the frame) at one end. In stock form they are more or less parallel to the ground, but if a taller spring is installed without lowering the radius arms' attachment point on the frame, the axle halves are forced to "roll" forward on their axis. This in turn reduces caster. Definitely the "spookier" of the two alignment specs at highway speeds, and aggravated by the fact that short radius arms change angle quickly when lift is added; getting the caster out of whack equates to a vehicle that is borderline undriveable. It does not take much to get caster out of proper specifications with TTB, and it does not take much caster change to make a normal vehicle handle like a nightmare. Throw some camber issues into the mix, and most would rather walk than

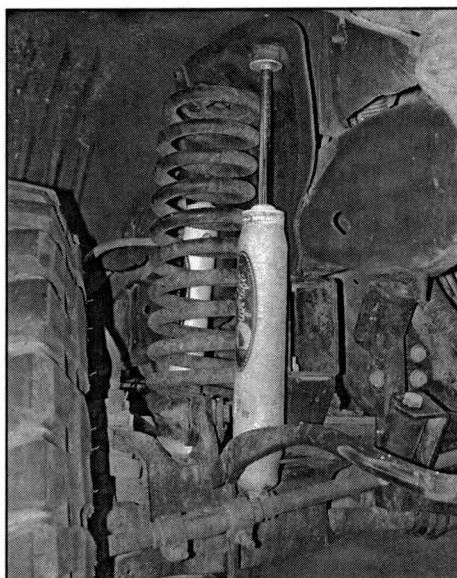

Spring height is the great wild card in TTB suspension designs because the springs play such a critical role in alignment. A spring that is just a little too tall or sagged just a little can have profound affects on the vehicle's ability to maintain alignment and related tire wear characteristics.

drive a vehicle with bad TTB. Radius arm drop brackets are the standard for correcting caster, with adjustable alignment cams enabling a tech to "fine-tune" an alignment.

Another highly recommended option is to replace the short factory radius arms with longer arms available from many lift companies. While an optional upgrade that can be expensive, extended radius arms fix several problems inherent to TTB. Specific to caster, properly built radius arms have caster correction built into their design while their longer length reduces the amount of caster change as the suspension cycles.

Spring Height

The wild card when it comes to alignment, spring height is what makes TTB so fickle. With other sus-

Quite possibly the most common mistake consumers make is underestimating the weight of aftermarket accessories and how that extra weight will impact the suspension. This winch and bumper combination weighs a conservative 160 pounds, but it's also mounted quite a distance from the front springs, which greatly amplifies the weight factor. With TTB, even a ½ inch of spring sag due to the extra weight here can cause camber issues that in turn creates irregular tire wear.

pension designs, one or at most two of these specs are impacted by the height of the springs, but the nature of TTB and its factory steering linkage throw all three into the mix. As a result, a spring-height change of as little as 1 inch can throw one or more of these specifications out of whack. Adding weight to the vehicle (via a heavy-duty bumper and winch or snow plow), using it hard off-road, and simple age can cause irregular tire wear and handling traits. For this reason, it is smart to record ride height at the time a lift is installed and then plan on replacing the springs or re-aligning the vehicle as the front springs settle.

Steering Straight

Take another look at the diagram

6" lift with stock steering linkage

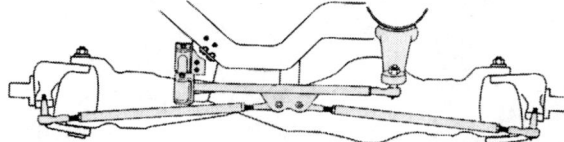

6" lift with dropped pitman arm and Superunner Steering System

This diagram clearly shows the difference between the stock linkage and the Superunner system. With the stock linkage and a pitman arm, the linkage is unable to follow the same arc of travel as the two axle halves, which results in poor handling characteristics. With the Superlift system, the tie rods operate at precisely the same angle as the axles. All hype aside, this system provides better steering characteristics than these trucks exhibited in stock form. The Superunner system only works with 4- to 6-inch lifts. (Courtesy Superlift)

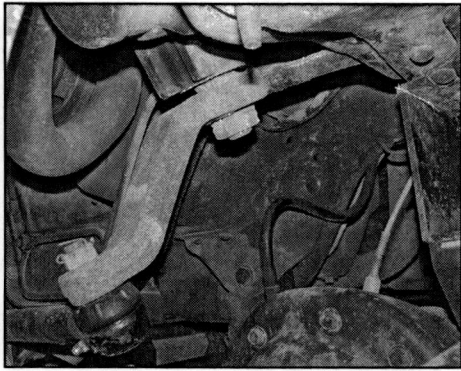

A drop pitman arm is absolutely critical with a TTB system in the 4- to 6-inch range. Some may find it interesting that TTB Ford (and several other Ford vehicles for that matter) pitman arms mount 180 degrees from most others, placing the drag link attachment point ahead of the steering box rather than behind.

at the beginning of this chapter and take special note of the steering linkage. In factory form, it is designed to have the linkage follow the movement (or arc of travel) of the two axle halves as they move through their respective travel cycle. While this system is adequate in stock form, even in unmodified form the common complaint is that the steering is vague and feels disconnected from the road. Unfortunately, due to its inherent design, changing the operating angle of the suspension above 2 inches throws the steering geometry way out of whack, meaning it no longer follows the same angle of movement as the axle halves. This in turn leads to all sorts of strange handling traits.

The correction common to all manufacturers is a dropped pitman arm, which does an adequate job of putting the steering back into its proper orientation with axle movement. With 2 inches of lift, generally no steering correction is needed pro-

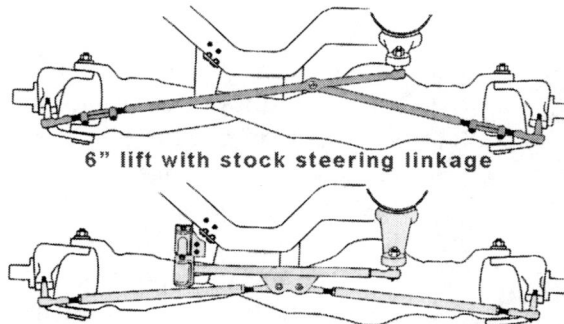

vided that the linkage is in tip-top condition and proper alignment bushings are used. A pitman arm is absolutely required on lifts between 4 and 6 inches regardless of the manufacturer. However, the vague handling characteristics evident even when stock are usually amplified with a lift system. Since the steering design does not lend itself to modification, what are the alternatives?

Enter Superlift's award-winning Superunner steering system. Truly revolutionary at the time it was introduced, the Superunner system wipes the slate clean and starts over with a completely different linkage system. It utilizes a centerlink that attaches to the pitman arm on one side and an idler arm at the other. Individual tie rods connect to the steering knuckles from the centerlink. This design, intended for 4- and 6-inch systems, addresses all of the shortcomings inherent to the factory linkage. Though optional with all of Superlift's systems, the Superunner steering can be easily added to its own brand of lift systems as well as most competitive 4- and 6-inch systems. Keep in mind that it may be necessary to purchase a Superlift dri-

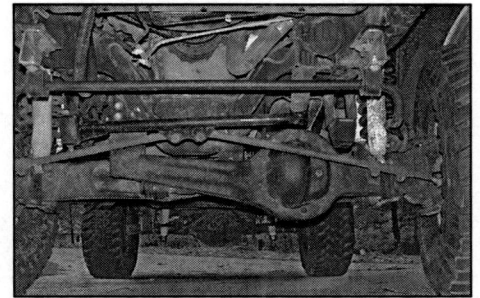

The Superunner system as installed on a TTB Bronco. These systems are also available for leaf-spring TTB trucks.

ver-side axle pivot bracket in addition to the linkage system depending on the design of the brackets of another lift brand. The advantages are obvious: a steering system designed with a lift in mind, and it truly performs as advertised (otherwise it would not have remained on the market for the last 15 years). The only drawback is that the system uses steering components that are proprietary and are not available at a parts store in the unlikely event of a failure in the field.

Tire Wear

One of the most common complaints of Ford owners from this era,

Inspecting the front tires on a TTB truck usually reveals whether or not it's time to take it back to the alignment shop.

whether the truck is two- or four-wheel drive, is tire wear. The problems are well documented when these vehicles are stock, and a lift only exacerbates matters. Most of the issues have to do with the components already identified, including axle pivot drop, radius arm drop, and spring height. The good news is that this tire wear can be controlled with adhering to a strict schedule of regular tire rotation and careful attention to both balance and tire pressure. Even so, it can be helpful to identify the type of irregular wear the truck is showing and how to fix it.

Inside Wear

If the inside of the tire is exhibiting more wear than the outside, then there is a problem with camber or not enough toe. Usually this can be diagnosed by simply looking at how the front suspension sits. If the bottom of the front tires appear to be tilted outward, then there is not enough camber due to either spring sag, improper axle pivot brackets, or it is simply out of alignment. Installing adjustable camber bushings and having the vehicle re-aligned and/or replacing the front springs can solve the problem.

Outside Wear

Though aggressive driving habits can lead to excessive outside wear (such as taking corners at speed), generally speaking the opposite is true of inside wear: too much lift has been installed for the axle pivots or too much toe is present. If the tops of the tires tilt outward when the vehicle is at rest on a flat surface, then too much camber is present. Reducing ride height via different coils, proper axle pivot brackets, and adjustable alignment bushings are all plausible solutions.

Cupping

If tires are cupping, meaning that individual tread blocks are wearing at different heights, then a simple alignment can usually solve the problem. Usually cupping is the result of two or more specs being mildly out of alignment. There are two important things to remember with cupping: the more aggressive the tread, the more likely cupping will exist, and once cupping starts, it is hard to stop. Sticking to a strict regimen of tire rotation and staying on top of alignment is important with any vehicle, but is especially so with a TTB Ford.

Radius Arms

The subject has been mentioned briefly, but radius arms on a TTB sus-

Though it has probably already been made perfectly clear, extended-length radius arms are the only way to go on an off-roader or a show truck. They look good and perform even better. This particular arm is manufactured by Superlift, but just about any company that makes a TTB lift system has extended radius arms available as an optional upgrade.

pension are more complex than they first appear. In factory form, coil-sprung TTB Fords use shorter versions of the radius arms found on previous-generation trucks. Constructed of stamped steel, they do their jobs adequately, but serious off-road use was obviously never in their design parameters. Stock TTB radius arms pose a double-whammy to off-road users: they introduce substantial caster changes as the suspension cycles and they inhibit readily available travel. Both problems are due to length. Though they serve the street user adequately, replacing them with aftermarket arms can offer substantial gains.

A typical aftermarket radius arm is upwards of 14 inches longer than the factory pieces. In addition to enhancements in looks, they offer the following advantages:

Caster correction: All of the necessary caster change is generally built into the arm itself, eliminating the need for drop brackets. This also

What Fits, What Hits — Ford

Model	Year	Additional Modification	31	32	33	34	35	36	37	38	40
Bronco	1980–'96	None		2	4		6				
		Fender Trim			2	4		6			
F-150	1980–'96	None		2	4		6				
		Fender Trim			2	4		6			
F-250	1980–'97	None					4	6			
		Fender Trim							4		6
Ranger 4WD (93–97 Mazda)	1983–'88	None	4	5½							
		Fender Trim		4	5½						
	1989–'97	None		4	5½						
		Fender Trim			4	5½					
	1998–'05	None			4						
Ranger R 2WD (93–97 Mazda)	1983–'88	None	5								
		Fender Trim		5							
	1989–'97	None	5								
		Fender Trim		5							
Bronco II	1984–'90	None	4½								
Excursion	2000–'03	None					3		5		
Explorer (93-94 Mazda Navajo)	1990–'94	None	4	5½							
		Fender Trim		4	5½						
	1996–'01	None			4						

Because the ability to maintain lift height is so important with TTB, choosing the right spring may mean the difference between being happy with a lift system and not. These springs are progressive, which can be identified by the more closely spaced wraps towards the top and wider spacing towards the bottom. This provides both good ride characteristics on the street and the extra rate needed to be good off-road performers.

improves ground clearance.

Caster change: The longer the arm controlling the axle, the less caster will change as the suspension cycles.

Suspension travel: As the radius arm gets longer there is less resistance from the bayonet bushing at the frame, so more travel is available to soak up the bumps.

Tire clearance: All of the aftermarket radius arms assume that a larger tire and wheel package is part of the picture, so the arms are contoured to accommodate this. Though tire-to-radius arm contact may still occur at a full-lock turn, there is still more turning radius available.

Appearance: Extended radius arms look cool. Usually tubular in construction, extended radius arms usually appear like the race designs that inspired them.

One important thing to keep in mind is that virtually all lift systems can be upgraded with longer arms after a lift is installed. Therefore, if budget prohibits doing the whole upgrade at once, these can be added later for better suspension performance. However, if the short arms are going to be retained, drop brackets are a must for any lift above 2 inches. One last thing, regardless of whether the stock radius arms remain or longer ones are installed, it's safe insurance to plan on replacing the radius arm bushings. Not usually included with standard lift systems, these are considered optional, but then again these vehicles are not getting any younger.

Choosing the Right Spring

It has already been covered but bears repeating: spring height is critical to maintaining proper alignment with TTB. Therefore, choosing the right spring can make or break a TTB suspension system. Most lift manufacturers also recognize that TTB is the beginning of most company's focus on ride quality; due to its design emphasis, lift development for TTB marked the first time ride quality was an equal concern to lift for many of the older companies. Though most have revised (often several times) their spring rates over the years for this era of vehicles, it is important to remember that a spring with excellent ride characteristics and superior longevity in its maintenance of lift height is the brass ring. Some companies err on the side of ride quality and end up losing height over time, while others focus on durability at the sacrifice of some ride quality. As a result, proper spring choice becomes a matter of research on what works best for those who use their vehicles for similar purposes and the suggestions of a manufacturer. One important thing to consider is that curb weight has a profound affect on how the vehicle rides and how long the springs will last. Adding an extra 150 pounds of weight for a bumper and winch may sag springs tuned for durability, while a topless and unloaded Bronco may have a harsh ride with springs designed around ultimate durability. It comes down to what makes sense for the individual application, but because springs impact alignment, it is important to make the right decision based on individual needs.

One tip to keep in mind when choosing the proper spring is that the ½-tons and the mid-sizes share the same spring dimensions and the springs are therefore interchangeable. Consequently, if a winch bumper on a Bronco is proving to sag the recommended springs, swapping to those from a heavier Super Cab truck may solve the problem. Select companies also offer shims for under the coil buckets on the axle. This corrects both mild-ride height issues and levels the truck if it leans slightly to one side. Though it may require some trial and error depending on how the vehicle is equipped, choosing the right spring could likely prevent frequent alignments or replacing springs on a regular basis.

The Bottom Line

Fords of this generation are well known for their problems, among them the ability to wear out tires quickly. Why? Because of its design, the simple fact is that just one thing being out of spec (such as springs that sag an inch) can impact the suspension in a major way. The inherent problem with TTB is that several components can cause both caster and camber changes to occur. The failure to recognize these components can lead to the infuriating issues that have caused TTB to have such a controversial past in the first place. Steering correction, though effective, still puts the system at the edge of its comfort zone and just one component that is excessively worn can wreak havoc. Unfortunately, the tire-wear problem is inherent to the design: significant caster and camber changes as the suspension cycles means more wear on the inside and outside edges of the tires on the vehicle in question. The bad part is that these vehicles are quite simply more difficult to achieve long tire life on than later models. By design, everything must be "ideal" for TTB to have good alignment and tire-wear characteristics; anything outside the ideal equals the traits people commonly complain about. In the end it should be recognized that the common complaints are known even when the vehicle is stock, and any equipment weaknesses a TTB Ford might have are magnified with a lift system, so it is important to start with a good foundation and/or be willing to invest the money required to bring a vehicle up to standards before a lift even comes into question. But a lifted TTB truck not only looks cool, it offers significant performance gains once the truck ventures off-highway.

TTB will go down in 4x4 history as either the most loved or hated suspension design. Often the underdog and always controversial, the fact of the matter is that TTB Fords are a little like older British sports cars: thoroughly enjoyable when they work, infuriating when they don't, and endearing when their attributes are understood in the midst of their quirks.

For the extreme crowd, this may be the ultimate answer: starting over with a solid front axle. Conversion kits are available from several sources, including Sky Manufacturing and Off Road Unlimited. Though medium-to-difficult fabrication is involved and easily installed donor axles can be hard to come by (like this high-pinion Dana 60), the deflated price of TTB trucks makes this a viable option for those off-roaders looking to benefit from late-model improvements like fuel injection. All it takes is a little know-how to swap a solid front axle in the place of TTB.

Tackling the TTB Beast

It has been hinted at elsewhere in this chapter, but let's quit beating around the bush: Ford TTB is the most universally hated 4x4 suspension system in history. It looks funny, acts funny, and has a dubious history with alignment shops. It's true that TTB is temperamental and has trouble staying within proper alignment specs, but Ford must have done something right because they sold millions of TTB-equipped trucks during their 16-year run. A whole bunch of those are still on the road today and can be had cheap, which makes them excellent candidates for weekend toys. And in the end, very few experts choose a late-model IFS system over TTB when it comes to pure off-road performance.

The truth is that lifting a TTB Ford is all about controlling its fickle ability to maintain alignment. Early on a small company called Superlift identified this key component, and its extensive line of TTB systems is what put them on the map. Though their products may have evolved over the years, Superlift's TTB line is still considered the standard by which all other comparable systems are judged. If you want it done right, you wouldn't go wrong to choose a Superlift kit.

Superlift, like many other lift manufacturers, offers two different levels of lift systems. Standard systems include just the basics to lift the truck properly, and have few bells and whistles. Though adequate for street duty, those who want to run the dunes and tackle a few trails should strongly consider upgrading the radius arms. Superlift's Superunner arms are substantially longer than stock and unlock gobs of hidden suspension travel. Steering is also marginal, and the only bolt-on "fix" is Superlift's Superunner steering system. Though these components add expense to the overall equation, there are few upgrades that can so dramatically improve on- and off-road handling.

To see what lifting a TTB truck is all about, we traveled to Off Road Unlimited in Burbank, California, and watched its technicians do the full Superlift Superunner treatment to an '89 Bronco. With all the options and accessories on this truck, installation took a day and a half with one professional mechanic. Do it in a driveway with an extra set of hands and you're looking at a long weekend. Speaking of which, installing a TTB kit is deceptively difficult and is not recommended for the average shade tree mechanic. The whole front suspension must be removed and the axle halves are very awkward to handle without the proper equipment. Factor in the very likely possibility that some frozen bolts will be found along the way, and spending the extra cash to have a professional do the job is a wise choice.

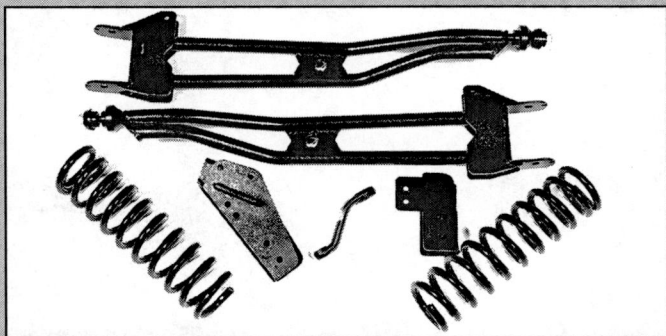

1. A TTB lift has lots of odds and ends, but the major players are all here. All TTB lifts above 2 inches have axle pivot brackets. Note that many manufacturers use the same brackets for 4- and 6-inch kits; there are simply two different axle pivot mounting holes present, so be sure to use the proper one for the lift. Coil springs provide the actual lift, a dropped pitman arm addresses the steering, and the Superunner radius arms address caster.

2. Installing the lift requires tearing the front suspension completely apart, so if this makes you uneasy, it's best to go to a professional. Techniques vary on securing the axle halves during the install. Some techs lower them to the ground, but this tech chose to strap them in place while installing the axle pivot brackets.

Tackling the TTB Beast *continued*

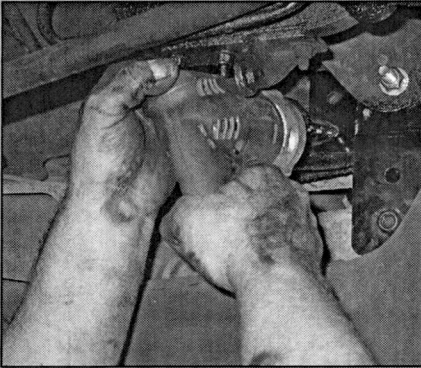

3. The first order of business is attaching the axle pivot brackets; these attach to the factory location on the engine crossmember. Like many other manufacturers, Superlift uses the same pivot brackets for both 4- and 6-inch systems; install the axle in the lower hole for 6-inch applications, and the upper for 4-inch kits. This is a 4-inch kit, but the technician temporarily attached the axle to the lower hole in order to keep it secure while the bracket was installed. Note two additional holes must be drilled in the crossmember.

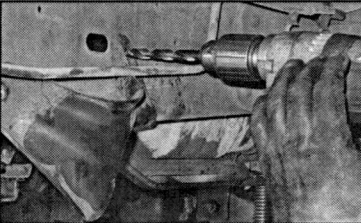

5. Now for the bad news: the factory radius arm brackets are riveted to the frame and must be removed. The rivets can be drilled out as shown here, or they can be very carefully torched off. With a standard lift system, a bracket supplied with the lift system spaces the factory mount straight down from the frame or it is replaced entirely.

7. After double-checking all measurements, the bracket was clamped in place and used as a template to mark and drill the new mounting holes. New Grade 8 hardware is used to attach the brackets to the frame. Always check for wires and plumbing on the other side of the frame before drilling.

4. The passenger axle pivot bracket installs just like the driver side. It uses two factory holes and then two additional holes are drilled for strength. Note the bracket also comes pre-drilled for the centerlink used in the Superunner steering system.

6. With the Superunner radius arms, it is necessary to measure back 18 inches from the factory location. This will be the new home of the factory radius arm bracket. In some cases there may be an interference problem with the transmission crossmember, but Superlift offers a simple relocation bracket to take care of this issue.

8. Use a jack to maneuver the axle halves into the pivot brackets, as they are both awkward and heavy. An extra set of hands is really necessary to do this safely. Note that the axle is now in the proper (upper) pivot hole for the intended 4-inch system.

Tackling the TTB Beast *continued*

9. Next up is installing the radius arms. Attach them to the frame brackets first, but pay close attention to the instructions for proper assembly of the bayonet bushings, as only one configuration is correct and a couple of the original bushing pieces are re-used. Once loosely attached to the frame, they can be attached to the axle halves with the factory hardware. If it looks like this procedure is a little funky, that's because it is; unless you do TTB Fords every day it's a struggle to keep everything properly supported and aligned during this process. The good news is that it's all downhill once the radius arms are bolted into place.

11. Most, if not all, coil-spring TTB trucks have accommodations for dual front shocks. However, not all kits come standard with four front shocks, so keep this in mind when shopping. Shock choices abound, but in this case the standard Superlift Superide shocks were selected. Any hydraulic shock from a reputable manufacturer will do, but if the truck is destined for high-speed stuff, remote-reservoir shocks should be considered.

10. Now that nothing heavy is teetering on a jack and at risk of falling, the installation can proceed with the coil springs. Be sure the coil pigtails are indexed properly, as some coil buckets have a recess built into them that allow each coil to seat properly. Reinstall the lower retaining cup and remember to install the factory rubber isolator on top of the coil before seating the coil in the upper mount.

12. After separating the drag link from the pitman arm and letting the factory linkage hang, the next step is bolting up the Superunner centerlink. The system comes with a pitman arm to replace the original as well as an idler arm that supports the other end of the centerlink. Designed to be used with 4 to 6 inches of lift, the Superunner system overcomes the excessive angles of the stock linkage when used with a lift. When tightening the system components, do not over-torque the idler arm bolts.

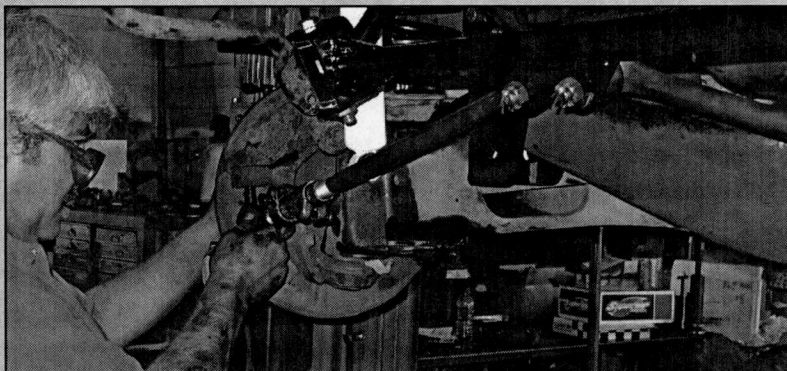

13. Next, the tie rods and adjusters are pre-assembled and roughly set to the proper length. The tie rods connect the centerlink to the knuckles. As you can see, the tie rods' angles match that of the axle halves. This eliminates the bumpsteer, wandering, and vague feel TTB is so infamous for. Final toe-in is set during alignment.

Tackling the TTB Beast *continued*

14. Not to be overlooked, proper operating angle of the sway bar is especially important for TTB's overall ride and handling. This is accomplished by lowering the bar's attachment points on the frame via the supplied C-shaped brackets. Disconnecting the bar greatly improves TTB's trail performance, but unfortunately no one makes a disconnect system at this time. However, trail-bound TTB owners can simply unbolt the sway-bar links at the trailhead... it really is worth the effort.

16. (Right) Steering stabilizers are particularly beneficial on a TTB truck, so adding a good-quality aftermarket unit is always a good idea. The Superunner steering requires the use of a specific stabilizer, but Superlift offers single and dual kits for the factory steering as well.

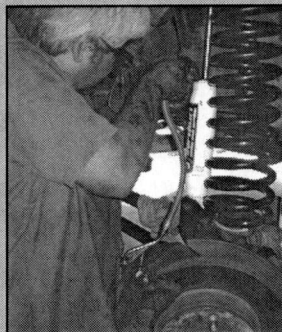

15. OE-style extended-length Superlift rubber brake hoses are the best choice if budget allows, but the factory hoses can be used with the relocation brackets supplied in the kit. Unlike conventional braided stainless hose, these lines include a leader at the caliper that directs the hose away from moving components and chafing hazards. They are also DOT approved.

17. The rear has the usual choice of lift options, but with this Bronco destined for the dirt, rear springs were the only real choice. The Bronco's short wheelbase makes it very sensitive to driveline vibration, so pay close attention to the instructions and be sure all degree shims are in place. In this case, the short factory block needed to be retained but the installers removed the degree shim after the test drive revealed a bit too much driveline correction. Also keep in mind that the Bronco CV joint is more prone to wear than normal, and a marginal one that had no vibration stock often starts vibrating after a lift is installed. Longer wheelbase trucks have less of an issue with vibration.

18. New shocks are a remove-and-replace procedure. Also note the sway-bar link on the frame. Though the rear sway bar worked fine with a 4-inch lift, another experience with a 6-inch system revealed longer links would be needed. In that situation, the installer ended up adapting a set of aftermarket links from a Super Duty to provide the necessary correction, as Superlift does not offer a bolt-in replacement. The factory links can also be cut and sleeved for more length.

Tackling the TTB Beast *continued*

19. (Right) Wrapping things up out back, an extended-length brake line is installed to match the front and the rear axle vent tub is also extended to work with the rear end's newfound extension travel. It's attention to little details like these that separate an average installation from a superior one.

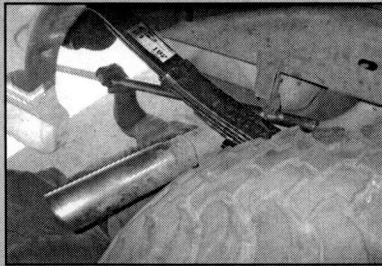

20. The installer encountered a common problem on this truck that is not often talked about: the increased arch of the rear springs caused the tailpipe to make contact with the passenger side spring (dual applications often hit both springs). Though light, this can make a lot of noise and eventually damage the exhaust system. In this case a little creative prying on the rear exhaust hanger netted the necessary clearance. Again, getting the little details right is important for a satisfactory overall finished product.

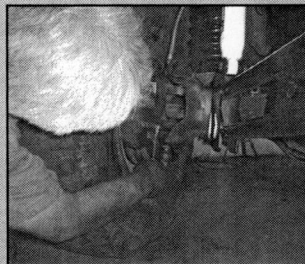

21. One last important adjustment is needed when using extended radius arms: the steering stops must be adjusted to prevent the tires from making contact with the radius arms. Tire-to-radius arm contact can be very dangerous. Other final steps included bleeding the brakes, double-checking all fasteners, re-aiming the headlights (another commonly overlooked procedure), and having the truck aligned.

We were ultimately unable to hook up with the owner of the Bronco for some finished shots and driving impressions, but we did find this white '92 Bronco with the identical lift but set at 6 inches rather than 4 inches. The Bronco provided an excellent ride on the pavement and a surprising amount of off-road prowess both in the rough and flexing its stuff on the trail. No more O.J. jokes for this Bronco!

All fixed up and ready to go, even TTB can be modified for better off-road performance—oh yeah, it looks better, too!

INDEPENDENT FRONT SUSPENSION (IFS) WITH TORSION BARS

This Chapter Includes:

'88–'06 Chevy Pickups
'92–'06 Chevy Blazer/Tahoe/Suburban
'84–'03 Chevy S-10 Blazer/Pickup
'97–'03 Ford F-150
'98–current Ford Ranger/Explorer/Explorer Sport Trac
'02–current Dodge Ram 1500
'03–current Hummer H2
'86–'93 Toyota Pickup/4-Runner
Various other import vehicles

High-speed stuff is not a problem for an IFS truck, making them ideal toys for playing in the dunes. (Courtesy Larry Conville/Skyjacker)

The 1980s represented a major change in the defining characteristics of a four-wheel-drive vehicle. Up to this point, 4x4s were thought of as primarily utilitarian; work trucks that needed the extra traction of four powered wheels to get the job done. While the recreational four-wheel-drive market began gathering steam (growing from a relatively small niche to a sizeable market segment) during this decade, the emphasis in four-wheel-drive design as the decade dawned was on durability and func-tionality. The bells and whistles stayed primarily with the cars, and a truck with leather seats was almost laughable. But the market was chang-ing and so too was the truck buying demographic. This new demographic wanted the safety, cargo-carrying capacity, and functionality of a truck, but they didn't want the "truck-like" ride to go along with it.

To attract these new truck buyers to their brands, for the first time ever truck designers were being asked to focus on ride quality. As popular and durable as the traditional sus-pension designs were, most of the truck manufacturers came to the conclusion that the only real "fix" for the ride quality issue was a major change in how front suspension was

The popularity of IFS Chevys represents a gradual shift in the role of 4x4s in the automotive market. They still serve as capable vehicles for getting to out-of-the-way places, but 4x4s these days see primarily street use and are more of a "support" vehicle for work and play. Still, a lift makes these trucks more versatile for a variety of purposes. (Courtesy Larry Conville/ Skyjacker)

their lives on paved roads. Nearly 20 years after the first mass-produced IFS 4x4s rolled off of the assembly lines it is clear that IFS is here to stay. Currently, only one light-duty vehicle (the Jeep Wrangler) is sold in the U.S. market with a solid axle. However, it is not all doom and gloom. Aftermarket suspension manufacturers have adapted well to the more complex IFS systems and have figured out how to enhance their performance substantially on the street and once the pavement ends. With 4x4s today sitting closer to the ground than ever before, adding ground clearance is the single biggest way to improve an IFS truck's off-road prowess.

constructed. Hence, four-wheel-drive independent front suspension (IFS) was born.

IFS is nothing new and has been around for decades in various forms under cars and two-wheel drive trucks. The benefits of IFS are readily apparent: the front suspension is essentially split in two, with each half able to react to the road surface without affecting the other side. This translates into improved ride quality and can be further tuned to improve cornering ability. IFS also allows the vehicle to sit lower to the ground, aiding vehicle entry and egress. But the problem facing designers was how to join the attributes of IFS with the necessary components needed to transfer engine torque to the front tires. The solution reached by all of the OE manufacturers (some sooner than others) was a control-arm based system with a rigidly mounted front differential.

While the ride quality improvements were substantial, to the traditional four-wheel-drive enthusiasts the gradual encroachment of IFS on

4x4s was the beginning of the end. While an IFS system has yet to be developed that has the same strength attributes as more traditional solid-axle designs, many argue that IFS is a more realistic approach to the role of a 4x4 because the vast majority of them spend more than 90 percent of

How It Works

With IFS, the front tires are attached to knuckles that pivot on ball joints just like solid-axle suspensions, but that is where the similarities end. Upper and lower control arms locate the wheel knuckles front-

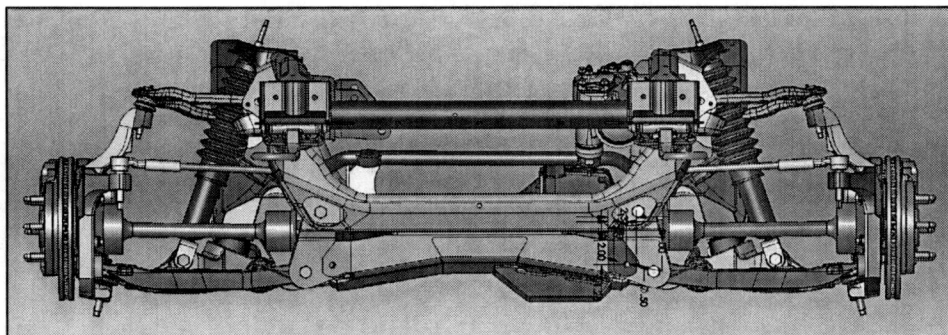

A typical lifted IFS configuration, in this case a Chevy ½-ton. (Courtesy Superlift)

to-back and side-to-side. The "spring" of the suspension is a torsion bar that attaches to one of the two control arms (usually the lower) and the frame. These torsion bars have a tremendous amount of preload or "twist" in them set by an adjuster at the frame end. The torsion bars are what determine the ride height of the suspension, and as the control arms move, more twist is introduced to the torsion bar, which has a certain amount of "give" that enables the control arms to move. The more the control arms move upward, the higher the torsion bar's resistance to twist, which is just like compressing a traditional leaf or coil spring.

The system that transfers power to the front tires is equally complex. With virtually all IFS designs, the front differential is rigidly mounted to the vehicle frame. Axleshafts with constant velocity (CV) joints on both ends move with the suspension and transfer power to the front tires.

The steering system is also complex, but can be easily understood. Instead of the usual tie rod and drag link arrangement, a centerlink attaches to a pitman arm at the steering box and an idler arm mounted to the opposite frame rail. Short tie rods then connect the centerlink to the steering knuckles. This design enables the tie rods to move in the same arc of travel as the control arms, which eliminates bumpsteer and other negative handling traits.

Suspension Basics

Unlike most other suspension designs, virtually all factory suspension components are re-used with a lift system. Instead of replacing components, lifting an IFS vehicle requires relocating nearly all of the

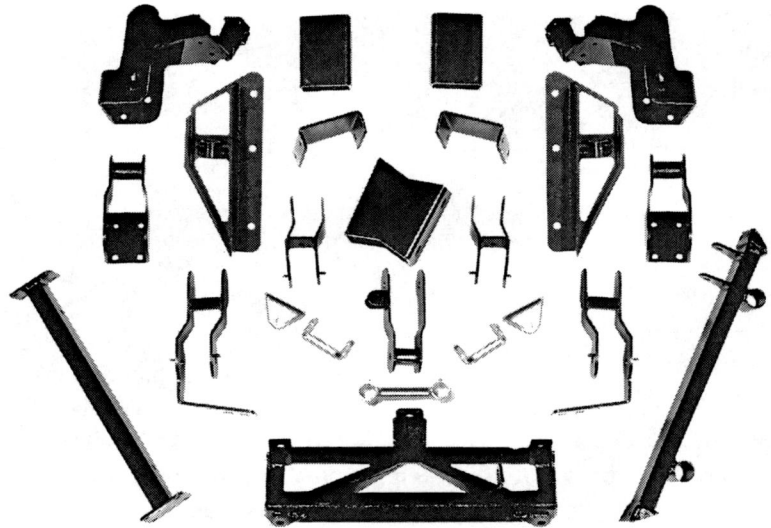

A typical bracket system consists of everything you see here. This system consists of upper control arm drop brackets, front and rear crossmembers, differential drop brackets, a centerlink, torsion bar drop brackets, sway-bar brackets, shock hoops, dual front shocks, and much more. (Courtesy Superlift)

front suspension components downward a certain distance, thereby achieving lift. This is true of any lift above about 2½ inches because of the narrow operating range of several key components. Since these systems are fairly complex, they are discussed in terms of the individual components that are impacted. However, before examining the components in detail, there are two different approaches taken with lift system design. Each has its pros and cons, and choosing the correct type of lift to best suit individual needs requires understanding each one.

The Bracket Method

The older of the two designs, a bracket-style lift system lowers *everything*. This is done via plate steel bracketry for the upper control arms and differential, while the lower control arms are typically treated to full-width crossmembers. Steering is typically addressed with a heavy-

duty replacement centerlink that lowers the attachment points for the tie rods. This design is a proven performer and enables the suspension to act just as it did in factory form, but with additional clearance for larger tires. There is no increase in front track width, so the front tires do not stick out past the fenders (assuming wheels with the proper backspacing are used), and the factory wheels (or wheels with the factory backspacing) can be retained. No differential modifications are usually needed, but some systems do require the use of a different front driveshaft (at additional cost). Bracket systems tend to be more complex to install than knuckle systems (and therefore more expensive), and some people claim to have problems with some systems being able to maintain alignment. These people often cite the source of the alignment problems to drop bracketry and crossmembers shifting when used hard off road.

A proven performer, bracket systems are a good choice for those who don't want to increase the truck's track width, or for those that live in areas where tires sticking out past fenders attract local law enforcement. However, this method can only be used on vehicles with the upper control arms mounted to the sides of the frame (such as Chevys). On other applications, including the Ford F-150 and late-model ½-ton Dodge Ram, the control arm is attached to the top of the frame. This position prohibits moving the control arm down from its factory location, making the knuckle method the only choice.

Knuckle Method

A more recent design, a knuckle kit allows the upper control arms to remain in their factory locations while still relocating the lower control arms downward. To span the increased distance between the two arms, a new

A bracket system re-uses all of the factory components in a lowered location, including the upper control arms. Though it's an older way to lift a truck, bracket systems have their merits and it's not necessary for a person that owns an older lifted truck to run out and buy a newer knuckle system. This truck even uses the factory wheels with larger tires, so the owner saved money by not having to buy new wheels.

The one complaint with bracket systems is that the additional brackets tend to shift over time, which can affect alignment and cause the occasional squeak or moan. Some shops "fix" this problem by welding certain components (like the crossmembers) together, but most experts agree properly torquing all of the bolts is enough to keep a bracket system aligned.

A knuckle kit is more difficult to design than a bracket system, but the end result is fewer parts and a cleaner overall look. In this view it is clear that the upper control arm stays where it was stock and the new knuckle spans the increased distance between the upper arm and the relocated lower arm.

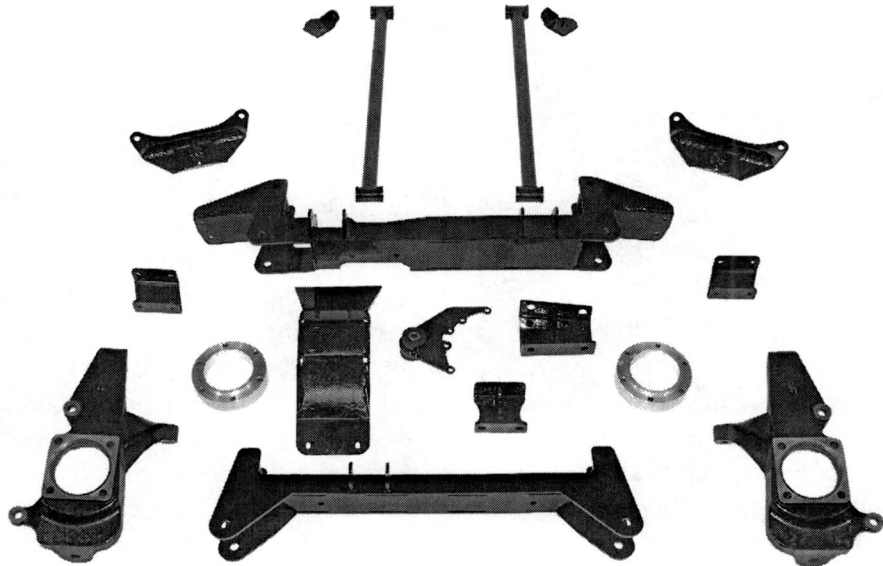

Knuckle kits require a few more special tools to install, but the installation itself goes faster than a comparable bracket system because the system has fewer parts. A typical knuckle system includes replacement steering knuckles, lower control arm crossmembers, differential drop brackets, CV axle spacers, torsion bar drop brackets, sway-bar links, and shocks. (Courtesy Superlift)

When considering a knuckle kit pay close attention to the wheel backspacing recommendations of the lift manufacturer. The longer neck of a lift knuckle usually interferes with a wheel that has the factory backspacing. The clearance is adequate here, but another ½ inch of backspacing would put the wheel (and tire) dangerously close to the knuckle. Always verify the proper backspacing before purchasing aftermarket wheels; because once tires have been mounted on them, wheels are not returnable.

Another drawback to reducing the amount of backspacing is the inevitable track width change. The amount of change varies among the different knuckle designs, but all knuckle systems change overall width. This causes the tires to stick out past the fenders, which means wider fender flares are needed to keep mud and other debris from being flung down the side of the truck.

cast-steel steering knuckle replaces the factory piece. This knuckle also incorporates steering correction by raising the tie-rod mounting-boss, so all of the factory steering components can be retained. This method greatly simplifies the kit and therefore the installation since several things are addressed within the knuckle's design. However, any cost saving gained by the simplified kit and installation are usually negated by the extra cost and machining time required for a replacement knuckle. In other words, bracket and knuckle kits are priced nearly the same.

At first glance the knuckle method appears to be the clear winner, but simplicity is not without its cost. The longer knuckle creates interference issues with factory wheel backspacing, so new wheels are usually required with the proper offset to clear the knuckle. Furthermore, most knuckle designs increase front track width (sometimes as much as 3 inches per side), so the front tires end up being farther apart than the rear tires. Though not much of an issue in terms of handling, increased front track width does push the tires past the edges of the fenders. This in turn requires additional coverage (via fender flares) or simply putting up with mud, rocks, and other debris being thrown down the sides of the truck while off-road. And while it is not often discussed, certain knuckle designs also quicken the steering ratio, change the Ackerman angle, and reduce turning radius. This can lead to some "funny" handling traits, not to mention making the truck more difficult to maneuver in tight parking lots. Somewhat related to the knuckle design, the front differential's pinion angle is usually corrected to allow

the front driveshaft to be retained, but the new position often requires trimming off a factory differential mount (on Chevys) or trimming some of the ribbing of the differential housing. These alterations can make it difficult to return the vehicle to stock (though select companies offer return-to-stock brackets).

Though they are a newer lift method, knuckle systems are rapidly becoming the standard for all IFS lift systems. Even with the track width changes, the simplicity of the knuckle design makes it an attractive choice with many of the same performance improvements as a comparable bracket system.

Control Arms

As mentioned earlier, the control arms are responsible for locating the steering knuckles. They also control alignment, with adjustable cam bolts providing the necessary adjustment at either the upper or lower arms where they attach to the frames. The lower control arms are always relocated, while moving the upper's relocation and/or replacement is dependent upon the style of lift and application. With most systems, the factory arms are retained and are more than capable of withstanding quite a bit of abuse. Select systems include or have replacement upper control arms available, which either correct suspension geometry or increase suspension travel (or both). Though these add expense to a lift system, generally, replacement upper control arms are the mark of a quality lift system where proper alignment and extension travel has been incorporated into the design. Avoid any system that calls for cutting or trimming the factory control arms

Differential drop brackets are normally employed to both lower the factory differential housing and alter its position to correct the front driveline angle. The amount of drop is critical to keeping the CV axles alive, and its rotation can mean the difference between an expensive replacement front driveshaft and being able to re-use the factory shaft. Note that many lift manufacturers do not recommend operating trucks equipped with full-time four-wheel-drive capability in that mode at speeds above 25 mph with their lift systems. If that is necessary, a replacement CV shaft is usually the solution.

With most knuckle systems the rotation of the differential housing (for driveshaft correction) and its new location means the upper mount on the driver side must be removed for steering clearance. This is easily accomplished (as shown), but makes the truck more difficult to return to stock. This kind of alteration may be objectionable to some truck owners, but a couple of lift companies have return-to-stock brackets to ease their fears. A new mount is bolted to the housing to replace the mount being cut off.

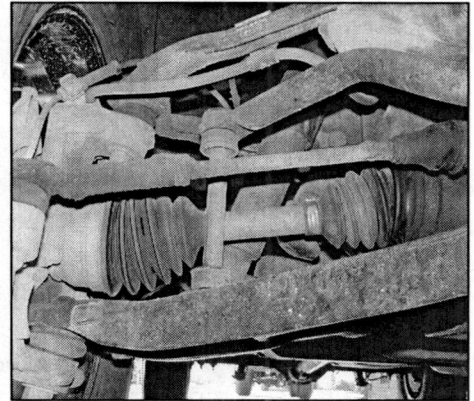

CV axle operating angle much beyond what can be seen on this stock truck means trouble. Even in stock form the boots rub against one another, as indicated by the clean areas of the boots. The critical part for all lift manufacturers is lowering the differential enough to be sure the CV axles are very close to these same angles at normal ride height.

Even though this CV axle is at full extension travel, this illustrates an important point. If the CVs are forced to operate at these angles all the time, failure is not far behind. Also, this photo is a reminder that the front suspension has a significant amount of movement; so if the axles look like this at ride height, imagine what they would be like at full extension. Also notice the aluminum spacer between the differential and CV, which compensates for the track width increase of the new knuckle.

(usually for shock clearance), as this weakens the arms and is a recipe for later failure.

Differential/CV Axles

It has been established that the differential housing is rigidly attached to the frame and CV axles are responsible for transferring power to the front tires. Though certainly effective, relocating the differential to work with a lift system must be spot-on or a multitude of problems develop in the front portion of the drivetrain. In addition, the CV axles are probably the most controversial components in IFS.

The differential must be lowered the same amount as the rest of the suspension system in order for the CV axles to remain within factory specifications. Moving the differential requires addressing front driveshaft angle, and with some systems a new front driveshaft is required (be sure to ask if the price of the driveshaft, if needed, is included in a price quote). The majority of IFS axlehousings are aluminum (another weakness compared to cast iron solid axles), but ring and pinion size is comparable to their solid-axle counterparts. With some lift systems (usually knuckle kits) it is necessary to trim the differential housing, which some truck owners may not like (after all, axlehousings are not cheap) and which can make returning the vehicle to stock more difficult.

As for the CV axles, there is not a lot of tolerance for error. CV axles are designed to operate within a very specific range of movement (the factory amount of suspension travel). If forced to operate outside its design parameters, the CV axle will wear and fail in a short period of time. One of the common problems with CV axles is ripping of the protective dust boot that keeps dirt out and grease inside the moving CV axle parts. Once again, CV boot failure is indicative of improper geometry or ride height that is set beyond the maximum recommended specifications. Another issue with CV axles is that they are not designed to have any inward or outward movement at the CV, yet it is known that the design of replacement knuckles in a few lift systems causes this movement to occur. The end result is the same, premature wear and failure.

Adding fuel to the fire, IFS differential components were designed around factory-size tires and wheels without much of a margin for strength beyond factory parameters. The same thing can be said for most other IFS components. A heavier tire and wheel package creates more rotating mass, which means more stress on already marginal differential and CV axles. Unfortunately, short of replacing the entire system with a solid axle (a viable but costly alternative that involves extensive custom fabrication), not a lot can be done to beef up these weaknesses beyond making sure a lift system has all of the necessary issues corrected in the design of the knuckles and drop brackets. One exception is Rough Country, which offers high-quality race-inspired CV axles designed to operate at greater angles than the factory pieces. Rough

One of the very few to address the shortcomings of Chevy CV axles is Rough Country, who now manufactures heavy-duty replacement CV shafts that incorporate a slip-spline within the axleshaft to accommodate more suspension travel. It also incorporates generally beefier CV components. (Courtesy Rough Country)

Remember that the vast majority of front differential housings are aluminum and therefore are very susceptible to damage from trail obstacles. A system that provides full skid plate coverage should be a strong selling point to someone that intends to use his or her truck off-road. A full-width skid plate also means a stronger overall subframe assembly. This particular Superlift skid plate is made out of ¼-inch mild steel, so it's more than enough to keep the obstacles at bay.

Country even incorporates slip joints to prevent movement within the CV sockets, which can destroy conventional CVs. Even with the addition of these high-quality shafts, it's still smart to understand the strength limitations of factory components and be judicious with the throttle.

Skid Plates

One area in which lift system designs vary widely is in regards to front differential protection. Some lifts incorporate an entire rigid subframe that protects the front differential (and other vulnerable front suspension components) with a steel subframe that covers the whole front-to-back and side-to-side. With systems that employ crossmembers, this heavy-duty skid plate also serves to triangulate the assembly as a whole, making one very strong unit unlikely to shift, bend, or break. Other systems employ a fairly basic ribbon of steel to protect the low-hanging centersection of the differential and at least tie the front and rear crossmembers together. These systems also usually employ lateral bars that extend back from the lift components to help amplify rigidity. If off-roading is the goal, it is a wise choice to select a system that has a full-width skid plate to ward off rocks and other obstacles. However, a street-bound truck can benefit from the cost savings of a simpler system that covers up the basics and leaves the rest alone. In select situations, a lift company offers a skid plate system that is optional to augment their standard offerings, giving consumers the choice to upgrade as budgets and off-road experience improves. The main factor to take into account is the quality and thickness of the steel constructing the skid plates: ⅜₆- and ¼-inch mild steel are capable of taking a beating, while lighter gauges and lesser steels may not be.

Torsion Bars

Torsion bars are essentially straightened-out coil springs, and just like a conventional spring they have the biggest impact on overall ride quality. With most IFS designs the bars engage the lower control arms, and because the arms are lowered with a lift system so too must the frame end of the bars. This is most often done via brackets that

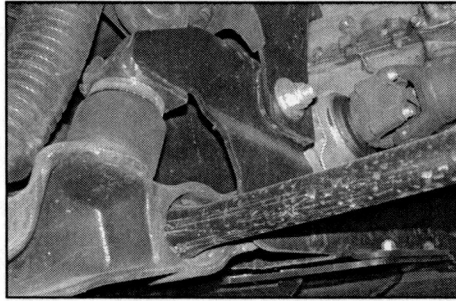

With most suspension types the torsion bars engage the lower control arms via a hex or splined joint. As the lower control arm moves, the torsion bar twists to allow deflection. Some designs (including Toyotas and Nissans) have the torsion bars attached to the upper control arm, and this is a plus since they can remain in their factory location with a lift and won't hamper ground clearance. As you might imagine, a tremendous amount of energy is stored in a loaded torsion bar and extreme caution should be used when doing suspension work on an IFS truck.

lower the torsion bar crossmember, but all of the factory components (including the adjusters) are retained. This lowered position hampers ground clearance where it is most critically needed (behind the front tires) and various attempts have been made to keep the bars in their factory location. Unfortunately, these attempts range from frighteningly unsafe to marginally successful. The main issue is that most designs place tremendous leverage on the torsion bars themselves, and this additional leverage forces the bars to sag and have difficulty maintaining ride height. Maybe someday a company will come up with the perfect design that allows torsion bars to remain in their factory location, but for now dealing with the ground clearance loss is the safest bet.

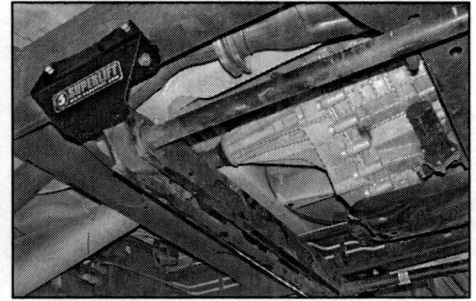

Lowering the torsion bar crossmember is a necessary evil when lifting most IFS systems. This crossmember serves as the frame-mounting mount for the torsion bars and is typically lowered the same amount as the lower control arms. Though it robs ground clearance, this is the only reliable way to enable proper operation of the torsion bars with a lift system.

The torsion bar adjustment on most full-size trucks looks like this: inside the crossmember is a "pork chop" that serves as a lever for the bar at the frame end. An adjustment bolt sets the amount of preload on the torsion bar; tighten the bolt to go up, loosen the bolt to go down. As you can see, this one has been maxed out due to a heavier tire and wheel package. It's always a good idea to measure the truck's ride height when it is aligned because the bars usually settle, and tightening the bolt can restore the lost height. Remember that fiddling with the adjustment for more ride height will impact alignment.

Torsion bars have an adjustment that enables the final ride height of the vehicle to be fine-tuned. This is why many IFS lift systems have a lift height range; the lift height depends upon where the torsion bars are set. This is handy because the bars can be adjusted up or down to accommodate slightly taller or shorter tires, and if the bars sag a little, a couple turns of an adjuster bolt brings the front of the truck back where it is supposed to be. However, there is not an unlimited amount of adjustment available and many misguided 4x4 owners simply crank up the bars until the adjustment stops. Though up to 2 inches of lift can be had in this manner, nothing destroys ride quality and expensive front-end components quicker. The many negatives of "cranking up" the torsion bars without a proper lift system is further discussed in the "Shorter Lift Systems" section of this chapter.

Leverage is the name of the game for a torsion bar, and lifting a vehicle can introduce unseen additional leverage. A heavier tire and wheel package places more leverage on the bars, as do wheels with reduced backspacing compared to factory. On heavy-duty ¾- and 1-ton trucks, the additional weight-carrying capacity of the factory bars can usually handle this additional weight and leverage, but lightly sprung ½-ton truck torsion bars may have difficulty maintaining proper ride height. Fortunately, there are a couple of tricks.

Quite often the compression travel stops for the front suspension are designed to be a part of the overall spring rate. This is especially true of late-model Chevy trucks, in which the bumpstops are touching (or nearly so) the control arms at normal ride height. The stops act as auxiliary

Compression stop clearance can have a big impact on ride quality. While the stock truck shown has quite a bit of distance between the compression stop and the lower control arm, with many IFS SUVs the stops are nearly touching at normal ride height. However, the lifted truck has quite a bit more clearance, which is good and bad. It's good because in theory this additional distance means additional suspension travel, but it's bad because the torsion bar has to carry the entire load of the truck. This can fatigue the bars and lead to sagging.

springs and assist the torsion bars by loading as the suspension compresses. If compression stop clearance is not addressed with the lift system, the bumpstops are rendered much less effective. This places additional stress on the torsion bars, which can in turn cause sagging and failure to maintain ride height. Always take it as a good sign when a suspension lift

uses the factory bumpstops and simply relocates them to work properly with a lift system, as the factory stops were designed with the torsion bars. Systems that don't address bumpstop clearance properly usually require adjusting up the torsion bars, which raises the spring rate.

Even with compression stop clearance taken into account, however,

With this Superlift system a high-quality rubber compression stop is utilized to "help out" the torsion bars. Further, stop clearance can be adjusted via spacers positioned between the stop and its mounting point. This enables the truck owner to fine-tune overall ride quality to suit his or her personal preference. This type of system is also more capable of handling high-speed dune running and helps keep torsion bar fatigue at bay.

Torsion bars often have an adhesive tag with a code that indicates its load rating, and it is also usually stamped in one or both ends of the bar. The vehicle dealer can decipher the codes for you and provide the codes for different models. Also note that torsion bars are side-specific, meaning they are built specifically for the driver or passenger side of the truck. Don't mix them up if you're installing a lift at home. Loading and unloading the torsion bars requires a special tool that can often be rented from the local parts store.

some vehicles are so lightly sprung that just a heavier tire and wheel package are too much for the factory bars. If this happens, there is still the commonly used option of swapping in bars with a heavier rating. Due to the surprising weight variances of different body styles within the same model, the spring rate of torsion bars are also body-specific. For example, if the torsion bars in a standard cab truck won't maintain ride height, swapping in a set of slightly heavier bars from an extended-cab or crew-cab can usually solve the problem. Though this may firm up the ride slightly, overall ride quality may actually *improve* with better cornering and less brake dive compared to torsion bars that are excessively cranked up. The opposite technique can be applied if the ride is too harsh. In the case of Chevy trucks,

Trim to Fit

Though virtually all suspension manufacturers claim a 6-inch lift will clear 35s and a 4-inch lift will clear 33s, what they don't tell you is that it often takes some light trimming to run the maximum tire diameter at a given lift height. For some reason the fenderwells on late-model trucks have more than enough height to accommodate a taller tire, but very little in the width department. Therefore, light trimming to the plastic valance may be necessary.

the torsion bar dimensions are the same from ½-tons all the way to 1-tons, so the sky is really the limit on tuning a ride to serve a specific purpose. Torsion bars are available from a few aftermarket sources as well as the dealer (even for older vehicles most of the time) and are fairly inexpensive. Conversely, some manufacturers include replacement torsion bars in their lift systems, and this is a good indicator that the company is concerned about ride height and the long-term ability to maintain proper ride height.

Steering

The majority of the torsion bar IFS systems utilize a conventional steering box with a centerlink and short tie rods as explained earlier. With bracket systems, the centerlink is either replaced or relocated with a lift system, while knuckle systems typically have all of the steering correction built right into the design of

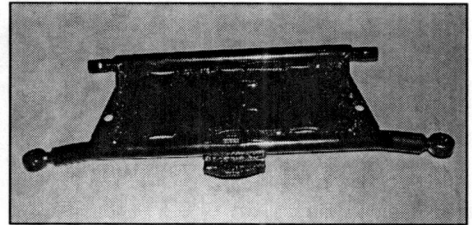

A factory centerlink is usually one bar that attaches to the steering and idler arms as well as the tie rods more or less in the same plane. IFS bracket systems include a complete replacement centerlink like this one. It is drilled to accept the factory steering and idler arms at the top, and it accommodates the factory tie-rod ends at the bottom. Also, an extra idler arm is usually employed to keep the centerlink from rocking fore and aft as the steering cycles.

the replacement knuckle. This simplifies installation and ensures that all of the factory steering linkage can be retained (and therefore replacement parts can be easily sourced).

It is not a coincidence that the tie rods tend to be short and roughly

With knuckle systems, all of the factory steering linkage is retained and kept in the factory location; the lift knuckle has a raised boss to accommodate the factory parts. It is also clear that the tie rod continues to operate at the same angle as the control arms and CV axles, which is important for ride and handling.

Idler arms, particularly those on Chevy trucks, have a dubious reputation for premature wear. This arm on a '99–2006 Chevy is average, but the idlers used on '88–'98 trucks as well as all S-10s are particularly bad. Known for wearing in as little as 20,000 miles, these idler arms should be checked frequently on lifted applications. Fortunately, Moog makes a heavy-duty replacement for full-size trucks that helps extend idler arm life substantially. Plan on replacing the idler arm on any high-mileage truck.

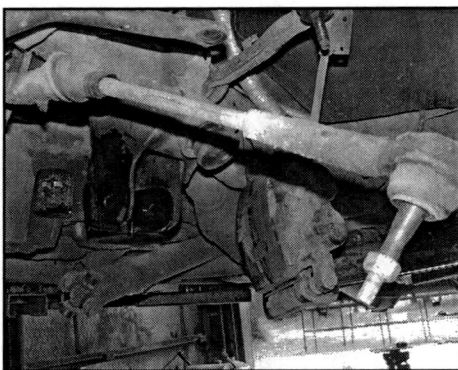

The tie rod ends are another common wear item on IFS trucks, and it seems as though all of them are marginal regardless of the make or model of the truck. Most everyone knows to check the outer tie rod, but did you know there's an inner as well? It's under that rubber boot on the other end of the link. Excessively worn tie rod ends can and have separated, causing a loss of steering control.

the same length as the control arms and run at the same angle between the centerlink and the knuckle. In order to ensure proper handling characteristics remain intact, any quality lift system will make sure that the tie rods follow the same arc of travel as the control arms.

Unfortunately, the steering system is also usually the weak point in some designs. Larger tires and wheels place a great deal of additional strain on the steering parts in particular, and this can often lead to premature wear and failure of several key components. For some reason idler arms are almost always problematic (on Chevys and early IFS Toyotas in particular) and it is not uncommon to find one excessively worn on a vehicle with as little as 50,000 miles in stock form. Because a bad idler arm can be a major source of several spooky handling traits, planning on a replacement (particularly on mid-size and '88–'98 Chevys) during a lift installation is inexpensive insurance against problems right out of the box. Inner and outer tie-rod ends can be just as problematic on all vehicles, so factoring in the extra cost of these items on a higher-mileage vehicle may help avoid nasty surprises during the actual installation.

Controlling Suspension Travel

Though often overlooked by both the end consumer and some lift manufacturers, controlling compression and extension travel is important with IFS due to the narrow operating range of tie rod ends, ball joints, CV axles, and so on. Extension travel control is built into a suspension's design in a variety of different ways, but the most common are

How the extension stops are addressed depends on the type of lift and the manufacturer. In this case a bracket is supplied and a urethane stop cushions the upper control arm (the factory stop is metal-on-metal). This particular truck is actually two-wheel drive, but many 4x4 versions are similar. This cushioned design is more conducive to off-road use.

either shock length or a frame contact point for the upper control arm. Since both can be impacted by a lift system, it is important to verify this has been addressed properly, as over-extending the suspension can cause major damage to many key and expensive components.

On the compression end of things it's a little bit different, but equally as important. Compression travel is usually controlled via a frame-mounted cushion that contacts the lower control arm. Not only do the compression stops keep the tires out of the fenders, they also have an important role in overall ride quality. This was discussed at length in the section on torsion bars, but in addition to their frequent use as "supplementary springs," they act as a cushion to absorb sharp blows encountered on and frequently off the road. Without them, sudden shocks would be harshly transferred to the passenger compartment, with potentially damaging results.

Suspension manufacturers address extension and compression travel a variety of different ways. Some relocate the factory stops via bracketry, while others may utilize replacement urethane stops. Each style has its pros and cons, but generally speaking, systems that utilize factory stops may offer a superior on-road ride, while replacement pieces may be more durable in an off-road environment.

Other Issues

Aside from weak tie-rod ends and idler arms, IFS systems have other weaknesses that should be known to any owner of an IFS truck whether it is stock or modified. It should be emphasized that these common weaknesses are not the *result* of a lift system, but are instead factory weaknesses that tend to be *aggravated by* a lift system and the heavier tire package that a lift allows. All of these weaknesses are well known to 4x4 shops and most are just as common to see on a stock vehicle as one that is modified. In select cases an aftermarket company (usually an OE replacement supplier such as Moog and its Problem Solver heavy-duty steering line for example) offers a heavy-duty version of a stock replacement part that is highly recommended when a lift system is in the equation.

Ball Joints

Light-duty IFS ball joints are infamous for premature wear, and it is practically uncommon to see a vehicle with over 50,000 miles on it with anything better than marginal ball joints. The upper joints are particularly susceptible to wear and even failure if not caught in a timely

Upper ball joints on IFS suspension systems are weak regardless of the vehicle manufacturer. Numerous things, including the weight of the tires and wheels, wheel offset, vehicle use, and maintenance all play a significant role in ball joint life. To make matters worse, some late-model vehicles like this Chevy do not have removable ball joints; the entire upper control arm must be replaced.

fashion. In fact, ½-ton ball joints are so weak and prone to premature wear that many well-respected suspension experts do not recommend anything taller than a 36-inch tire, with even more conservative recommendations for mid-size vehicles. To make matters worse, quite a few of the late-model vehicles do not utilize replaceable upper joints; instead the whole upper control arm has to be replaced, which adds to the expense. It is strongly recommended that the ball joints be checked frequently on a lifted vehicle and after every off-road excursion. Remember that ball joint failure can be sudden and cause a loss of vehicle control, so the importance of staying on top of the ball joints cannot be overstated! Ball joint life can be maximized by performing regular service, staying within recommended ride height, and choosing a wheel backspacing that is as close to the factory spacing as possible.

Wheel Bearings

Somewhere along the line a genius deep inside each OE manufacturer decided serviceable wheel bearings were not needed on a 4x4, and a unit bearing would suffice instead. With only a couple of import trucks as notable exceptions, just about every IFS truck uses a non-serviceable wheel bearing assembly commonly referred to as a unit bearing. Though it can be argued the design offers better sealing capability from water and other contaminants, history and common sense says bearings that can be serviced regularly will last for nearly the full lift of the vehicle. Service records aside, the real killer here is the price tag for a replacement unit bearing, which can run into hundreds of dollars *just for the replacement*

Late-model vehicles are equipped with unit bearings like this one that are not serviceable, so when they go bad, the entire assembly must be replaced. Unit bearings on a stock truck can go 100,000 miles depending on use, but heavier tires and wheels can take their toll on these pieces and reduce life by as much as a third. It's important to check the bearings on a higher mileage vehicle during a lift installation, as any slack in the bearings produces an unsatisfactory finished product when lifted.

piece. By comparison, a complete set (as in both sides) of replacement bearings for a common serviceable front axle can run about $150.

A larger tire and wheel package obviously has an impact on the wheel bearing, as does frequently overloading the vehicle. Frequent submersions and exposure to sandy environments also have an impact, as no seal is impregnable. Unfortunately, there is not a lot that can be done to prolong wheel bearing life beyond avoiding excessively wide tires and wheels with very little backspacing. Also, in most cases unit bearing replacement is a fairly easy procedure, but does require separating the stub shaft from the bearing, which in many cases requires removing a very large (and oddly sized) nut with a great deal of torque.

CV Axles

There are actually two types of CV axle failure: the failure of the protective rubber boot, and the failure of internal components. While the former can eventually cause the latter, ripping a CV boot is a very common complaint (although it seems to be less of a problem on newer vehicles). The actual failure of the CV axle is usually due to excessive throttle being applied with the steering at full lock, or a sudden shock-load to the axle while in four-wheel drive.

Assuming the lift system has been designed and installed properly on the vehicle, CV axle failure can most often be attributed to exceeding the maximum recommended ride height and forcing it to operate outside of its (pretty narrow) range. Though simple wear and tear can cause a boot failure, frequent failure on the same vehicle can usually be sourced to either kit design or exces-

Though they might look cool, lift systems in the 10- to 14-inch range just aren't practical or safe for a daily driver. There is way too much stress on already marginal components, and the 42-inch tires on this truck aren't helping matters at all. Though this might be okay for show, if major lift is desired on an IFS truck, the best option is a solid-axle conversion.

sive ride height. As for mechanical failure, once again this depends on how the vehicle is used. Judicious use of the throttle, especially in full-lock turns when the suspension is flexed, can usually mitigate catastrophic failures. Simply recognizing the weak link is the best defense against failure.

"Big Lift" Systems

There has been a trend in recent years for some companies to offer lifts above 6 to 7 inches for ½-ton vehicles. Though theoretically possible, it is not a coincidence that these companies seem to raise and disappear with the tides. The problem, once again, has to do with factory components. It is simply too much to ask a ½-ton ball joint to withstand the pressure of 38 inches (and larger) tires... many problems and cata-

strophic failure are the inevitable result. Generally speaking, a reputable lift company will "take" what the factory gives them and maximize its potential, yet for some reason there are companies always willing to push the envelope. Because a lift system relies upon factory components to survive, at some point a line must be drawn in the name of safety. Simply put, any lift above 7 inches (and that is a generous limit) is not safe on a ½-ton vehicle and should be avoided at all costs on a street-bound vehicle. Even if these systems address some of the factory shortcomings (such as ball joints and tie rod ends), there are many other factory components that are not sufficient to handle the daily stress of larger tires on the street, let alone in an off-road environment. Though these systems may be acceptable in a strictly "show" environment, an IFS truck with a huge lift should not be relied upon for daily driving and driving in general. In order to lift a ½-ton properly at taller levels, extensive fabrication and sufficient drivetrain beefing is needed in order to be considered safe. Step up to a ¾- or 1-ton and the margin improves slightly, but 7 inches should still be considered the maximum. Generally speaking, avoid 9- to 12-inch IFS systems at all costs.

Shorter Lift Systems

So far this chapter has focused primarily on lifts in the 4- to 6-inch range, but what about those people who want a little boost in order to step up a tire size or two? It has been mentioned earlier but bears mentioning again: generally speaking, an IFS system sits low in the front. Those simply looking to eliminate this factory nose-down rake have a

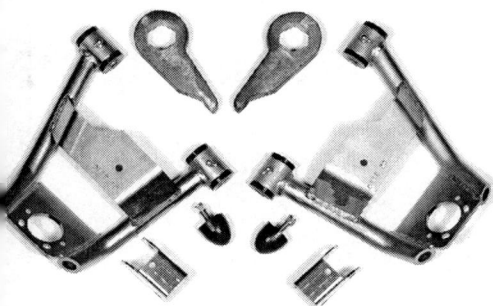

Two-inch lift systems are available for most popular vehicles, with the majority of them consisting of new upper control arms that address both extension travel and alignment. Some may also include torsion bars with a slightly heavier rating or re-indexed "pork chops" that enable more preload on the factory bars. (Courtesy Superlift)

Many people don't want the hassle of getting in and out of a truck with a 6-inch lift; instead, they simply want to level the stance of the truck and add a slightly bigger tire. 2- to 3-inch systems like this one accomplish this goal and provide a "lightly modified" look.

delicate balancing act ahead in that "cranking up" the torsion bars alone is not sufficient beyond about 1 inch of lift. This is because of the narrow operating window of key components that have already been discussed in detail. That said, it is possible to gain mild amounts of lift without breaking the bank and causing severe wear and premature failure to components.

At about 1 inch of lift, all that's needed is running up the adjustment to the torsion bars a little and re-aligning the vehicle. At taller lift heights, more is needed. Because of the OE fixation ono ride quality, it is not uncommon to see the existing torsion bars "maxed out" with no more adjustment possible at more than about an inch. Even though a heavier torsion bar may be available to run ride height further (and some "short" lifts provide replacement torsion bars), there are other issues at hand. First, there is extension travel.

If nothing else is done to the suspension but cranking up the bars, then all that's really happening is sacrificing extension travel for ride height. With some systems, a 2-inch lift places the extension travel at or within ½ inch of the extension stops. This means that the suspension only has to travel another ½ inch before being locked out by the factory stop. Since it's easy to encounter that ½-inch of travel under normal driving conditions, the suspension constantly "tops out" or hits that maximum amount of travel. This constant topping out can lead to a very harsh ride and accelerated wear of already marginal components like ball joints and tie-rod ends. Furthermore, increasing ride height alone puts alignment adjustments at their maximum range (which results in marginal alignment and excessive tire wear) and CV axles at their maximum acceptable operating range.

The solution for most quality-conscious lift system manufacturers is to combine a mild boost with items that address the shortfalls. If the factory design allows, sometimes only new torsion bars are required along with low-profile extension stops and other odds and ends. However, most include replacement upper control arms that both address alignment and extension travel, and drop brackets for the differential to alleviate CV axle angles. Once again, lift manufacturers will "take" what the factory gives them and address the rest as necessary. Some even include replacement torsion bars that bump up spring rate slightly, but keep in mind these are usually factory bars for a heavier application as discussed in the torsion bar section. Since short lift systems vary widely in what they include among lift systems for the same make, quality research on each is the best way to identify a system that best serves individual needs.

What Fits, What Hits

Model	Year	Additional Modification	31	32	33	34	35	36	37	38	40
CHEVY / GMC											
Avalanche ½-Ton	2001–'06	None		2	2¾		6				
Avalanche ¾-Ton	2001–'06	None					6				
Pickup ½-Ton K1500	1988–'98	None		2½		4	6				
		Fender Trim			2½		4	6			
	1999–'06	None	2		3½		6				
Pickup K1500HD	2001–'06	None					6	7			
Pickup ¾-Ton (6-Lug Wheel)	1988–'98	None		2½		4	6				
		Fender Trim			2½		4	6			
Pickup ¾-Ton K2500 (8-Lug Wheel)	1988–'98	None					5	7			
	1999–'06	None					6–7				
Pickup K2500HD	1999–'06	None					6		7½		
Pickup 1 Ton K3500	1988–'98	None					5	7			
	2001–'06	None					6		7½		
Hummer H2	2001–'05	None							6		
S-10 / 15	1983–'03	None	2		6						
S-10 / 15 ZR-2	1998–'04	None		2	6						
Tahoe / Yukon, Blazer	1992–'99	None		2½		4	6				
Suburban ½-Ton	2000–'06	None			2½		6				
Suburban ¾-Ton	1992–'99	None					5–7				
	2000–'06	None					6–7				
DODGE											
1/2-Ton	2002–'05	None					4	5½			
		Fender Trim						4			
FORD											
F-150	1997–'03	None				4	5				
NISSAN											
Hard-Body 4WD	1983½–'97	None	2½								
Hard-Body 2WD	1986½–'97	None	2								
TOYOTA											
4-Runner	1986–'89	None			4	5					
		Fender Trim				4	5				
	1990–'95	None			4	5					
		Fender Trim				4	5				
Pickup	1986–'96	None				4					
		Fender Trim				4					

Heavy-Duty Attitude

Heavy-duty Chevys are a popular choice for heavy haulers. They have powerful gas or diesel power-plants, the venerable Allison transmission, and a host of stout axles. They even sit a little taller than their ½-ton counterparts, which is never a bad thing. However, their stance is still not enough for getting into muddy construction sites or that remote camp spot that's the perfect jump-off point for a weekend of fun with the quads. No, if you want to utilize an HD to its fullest potential, there's gotta be some extra room under the frame to clear rough terrain, and some more room for bigger meats.

Fortunately, a number of lift manufacturers have heard the cries of HD owners wanting a taller truck, and a host of lift systems are available for this popular platform. One of the only manufacturers to offer two different lift methods, however, is Superlift. Since the new body style's introduction, Superlift has developed a 7-inch bracket system that does not increase front track width, and a more installer-friendly 6-inch knuckle kit. No matter what your priorities, either system can deliver the goods.

Lifting an IFS truck is complicated and something best left to the professionals unless a great deal of tools and ideally a vehicle hoist are at your disposal. The entire front suspension must be torn down and then put back together using the supplied lift components. However, don't let the complexity scare you into thinking that the truck will never be the same, because nothing could be further from the truth. Instead, a lifted HD acts just like it did stock because the factory springs and torsion bars are retained. As a result the handling is just like it was before, just with additional ride height.

To get the low-down on lifting an HD, we traveled to Superlift's R&D department as they did a final check-fit on their company's new knuckle system. All of the basic steps are the same regardless of which emblem is on the nose of the IFS truck in question, and in fact ½-ton Chevys are almost identical (just with lighter-duty factory parts). The install took a full day with two pros going at it, so expect substantially more time needed in the off chance you want to attempt the installation yourself. In the end, the truck cleared 35-inch tires with ease and now stands tall in parking lots full of lesser trucks. Best of all, load-carrying capacity is unaffected, but just make sure your gooseneck trailer has enough adjustment for the taller truck.

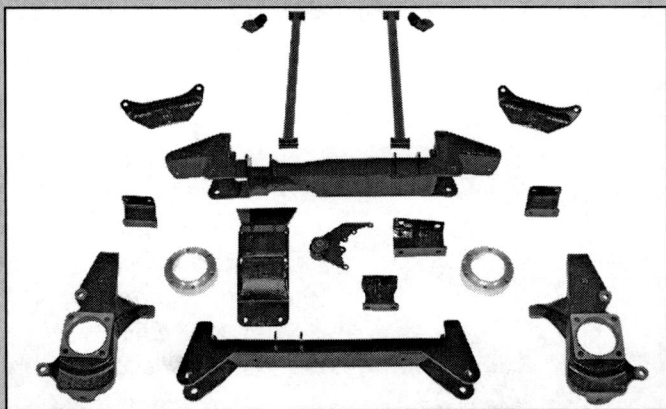

Superlift's knuckle system includes everything needed to lift an HD properly. This includes plate steel crossmembers, a skid plate, differential drop brackets, a differential skid plate, sway-bar brackets, CV axle spacers, and cast nodular iron knuckles. Not shown are the rear shocks and your choice of three different shock options.

Bigger, better, and badder attitude; complicated IFS is no match for a quality lift that provides substantial off-road performance gains.

Heavy-Duty Attitude *continued*

1. The process starts with tearing down the majority of the front suspension. This includes removing the differential, but the upper control arms can remain in place. The system requires trimming the driver side lower differential mount off of the frame. After removing the undercoating from the area, the technicians marked a trim line before breaking out the plasma cutter. Trimming this piece is usually required with any lift system.

2. A plasma cutter or torch makes short work of the required trimming. By removing the undercoating first, you avoid mild flare-ups and the scalding hot drips of liquid coating. It's also easier on the plasma cutter.

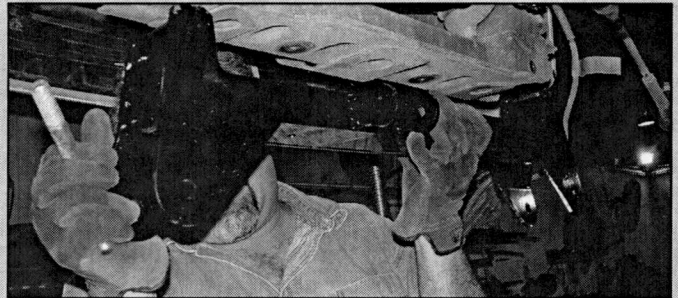

3. Superlift supplies a plate to reinforce the cut area that must be welded in place. The clearance between the differential (installed later) and the rear crossmember is tight so it's important to position the plate exactly as the instructions indicate. Here, Scott temporarily installed the crossmember and then tack-welded the plate in place to be sure an adequate amount of clearance was achieved. The crossmember was then removed for final welding.

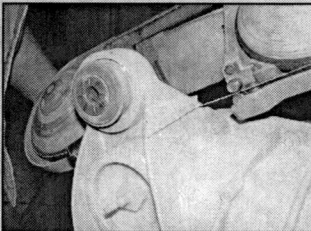

4. Tech tip: installing the steering stabilizer now (with the differential out of the way) saves a lot of headache later, as installing the stabilizer after the lift is complete can be difficult. Next, comes hanging the front crossmember, which attaches to the original lower control arm mounting holes.

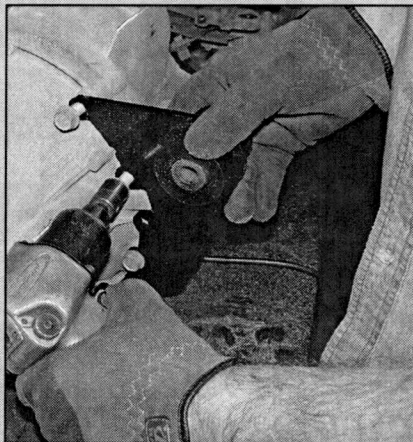

5. Before the differential can be installed, the driver side upper mounting ear must be trimmed off for centerlink clearance in the diff's new location. This must be done with a reciprocating saw, a cut-off wheel, a hacksaw, or a portable band saw. Never, ever use a heat source to cut the aluminum housing or you will ruin it. Yes, differential housings are expensive.

6. To replace the trimmed-off mount, Superlift supplies a bracket that attaches to the differential via existing bolts. The mount utilizes urethane bushings that should be lubricated prior to assembly. Be sure to tighten the bolts to the proper specifications using a torque wrench.

Heavy-Duty Attitude *continued*

7. Next comes raising the differential into its new home. It's a good idea to have some help for this part, as the housing is both awkward and heavy. The new upper mount attaches to the tabs present on the front crossmember, while the passenger-side mount is lowered via a plate steel bracket. These pieces lower the differential to work with the lift system and also "roll" the housing to correct the front driveshaft angle. The front driveshaft does not need to be replaced with this system.

8. With the differential in position the rear crossmember can be wrestled into place. It too attaches to the factory LCA (lower control arm) holes using the provided hardware. The rear crossmember also serves as the driver-side lower differential mount, and it can be a little tricky getting the differential to line up just right. This is why it's a good idea to keep the differential loose until all three bolts are installed.

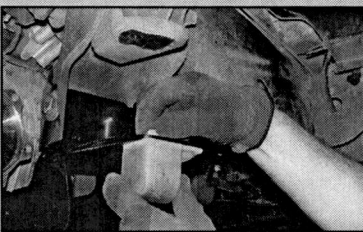

9. The factory compression stops play a large part in the overall spring rate of Chevy torsion bar systems and it's important for the stops to be positioned properly in relation to the lower control arms. The rear crossmember relocates the compression stops the required amount and one nut secures them in place.

10. Superlift provides a skid plate to protect the fragile aluminum differential. It installs between the two crossmembers and helps tie the two together for a stronger overall assembly. It even attaches with button-head bolts for a cleaner look.

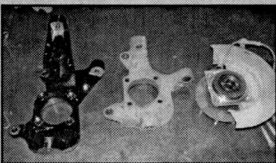

11. Tackling the knuckle assembly is easy, but care should be taken to make sure everything is installed in the proper orientation. The replacement knuckle (left) has a longer neck to accommodate the greater distance between the two control arms and also has a raised tie-rod boss for steering correction. The factory bearing assembly (right) needs to be installed in exactly the same orientation as it was on the original knuckle. However, the factory dust shield should be discarded. Use Loctite on the bearing bolts.

12. Once the knuckles are assembled, hang the lower control arms on the crossmembers and then attach the knuckle assemblies to the ball joints. Slide the CV axle into the hub assembly and then secure it to the hub using the factory hardware.

Heavy-Duty Attitude *continued*

13. The new knuckles push the front track width about 2 inches per side (for upper ball joint clearance), so to make up the difference Superlift supplies machined aluminum CV axle spacers that install between the differential and CV flanges. New longer hardware is also supplied, but a word of caution: Do not use an impact to tighten the CV axle bolts; they are easy to over-tighten and strip.

15. With the main portions of the kit in place it's a matter of buttoning up a few details. Be sure the ABS wire and brake hoses are secured as shown and routed per the instructions. This is also the time to install the shocks, sway-bar drop brackets and links, and hook up the tie rod. Leaving the tie rod end for last allows you to move the knuckle for better access to these components.

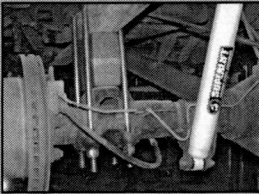

17. Rear lift is much simpler than the front and largely consists of installing the rear blocks and shocks. However, the rear brake hose also needs to be relocated from the top of the frame to the bottom. On this truck, the standard Superlift Superide shocks were utilized, but monotube gas and remote reservoir shocks are also available.

14. The kicker braces are lateral supports that provide the new subframe with extra support. They attach to the back of the Superlift rear crossmember and existing holes in the transmission crossmember.

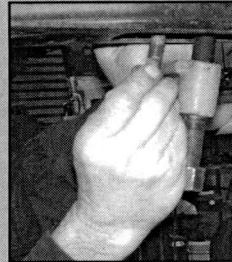

16. The last step for the front suspension is installing the torsion bar crossmember brackets. These attach to the sides and bottom of the frame directly below the crossmember's factory location. After the torsion bar crossmember is secure, position the torsion bars in the lower control arms, slide the adjuster arms into the crossmember, and slide the bars back through the crossmember and into the arms. Load the torsion bars with the same special tool required for removal and insert the nut block as shown. Run the adjuster bolt about two-thirds of the way in; final ride height adjustments are done once the vehicle is back on the ground.

18. A small but important part of the equation; be sure to install the supplied rear compression stop brackets, which simply space down the factory stops. Emergency brake hanger extensions are also supplied so that the E-brake cables remain properly routed with the rear axle's new location.

The finished suspension is clean and simple. Not a lot of bells and whistles but definitely well-engineered to hold up for the life of the truck. To add a little extra bling, the owner chose to add the optional stainless steel skid plate.

INDEPENDENT FRONT SUSPENSION WITH STRUTS

This Chapter Includes:

2005 and Newer Ford F-150
2007 and Newer Chevy
    Silverado/Tahoe/Suburban
2002 and Newer Dodge Ram 1500
1995 and Newer Toyota
    Tacoma/Tundra/FJ Cruiser
2001 and Newer Jeep Liberty (KJ)
2005 and Newer Jeep Grand
    Cherokee (WK) and Comman-
    der (XK)
2005 and Newer Nissan Titan

As evidenced by the latest crop of ½-ton trucks and four-wheel-drive SUVs, IFS with struts is the wave of the future. Ford began using struts on its four-wheel-drive light trucks with the introduction of the 2005 F-150.

As four-wheel-drive vehicles continue to evolve and appeal to a wider market, so too must the suspension. With just one or two exceptions, OE engineers design suspension systems today with excellent ride-quality as the ultimate goal. Unfortunately, ride-quality concerns and strength considerations are often at odds, with ride-quality usually winning out. However, that is not to say this latest crop of vehicles are not without their off-road performance traits, with virtually all of them better able to handle high-speed off-roading than their ancestors; it simply remains to be seen for how long they will be able to do so.

Suspension Basics

The latest generation of four-wheel-drive suspension design carries the same basic configuration as the previous IFS systems discussed in Chapter 6. Upper and lower control arms locate a steering knuckle front-to-back and side-to-side, while a rigidly mounted front differential transfers power to the front wheels

As trucks evolve and continue to become more car-like, they are gradually getting lower to the ground as well. The 2007 Chevy is a perfect example of this trend. A real detriment to off-road use, however, this lower stance does aid vehicle entry and egress. This, combined with new traction control systems, presents the latest challenges to aftermarket lift companies.

The 2007 Chevy ½-ton IFS system utilizes struts. Other strut-equipped vehicles are similar in design to the Chevy. (Courtesy Superlift)

Toyota started the trend in 1997 with the new Tacoma, but few guessed what a big role struts would play in future 4x4s. Coil-over struts are more compact than the torsion bars they replaced and offer more precise tuning for ride quality, hence they are quickly becoming the latest trend in suspension design. (Courtesy Skyjacker)

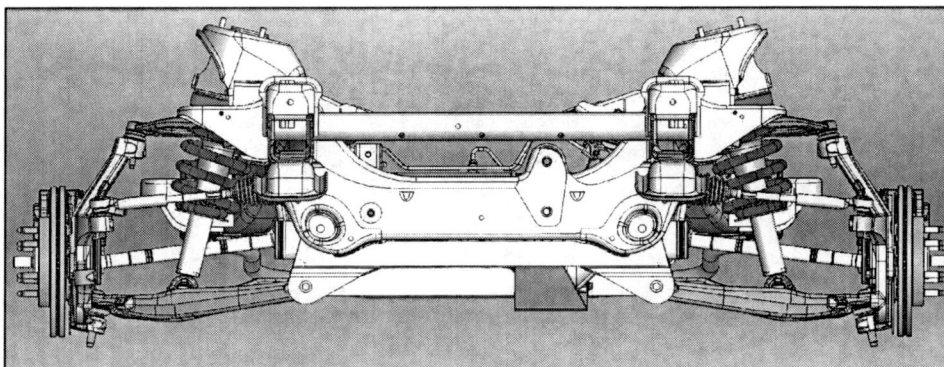

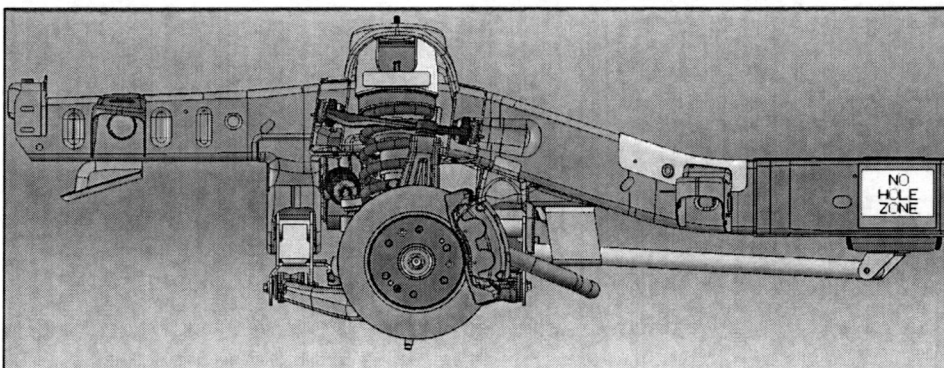

via CV axleshafts. Due to the design similarities, these new-generation IFS systems carry many of the same attributes and limitations as those already discussed in the last chapter. However, there are two major changes that have taken place, which remain the focus of this chapter.

Instead of torsion bars, these vehicles utilize a compact coil-over shock assembly to handle suspending and dampening duties. Though by definition these assemblies qualify as a coil-over shock, they should not be mistaken for the high-performance racing variety found on purpose-built off-road trucks and buggies. Instead they more closely resemble a strut, which has been used in cars for decades. Aftermarket suspension manufacturers use both terms. These coil-over assemblies allow more precise tuning for optimum ride-quality in a small, compact package.

The other major change has to do with steering. Up until the last few years all trucks utilized a traditional steering box and linkage system, which offers simplicity and superior strength. However, in the name of more precise handling,

Rack-and-pinion steering systems (the large assembly with black boots on either end) are the other new trend in light-duty trucks. Capable of more refined steering characteristics than the traditional steering boxes they replaced, rack-and-pinion steering is nevertheless bulky and very expensive to replace when it goes out. The rack is usually buried up above the differential, which makes it difficult to see without the front suspension torn down as it is here. This, plus the rack's size, makes it impractical to relocate when installing a lift. As a result, IFS strut systems are all "knuckle" systems with steering correction built into a replacement knuckle.

Lift manufacturers take a similar approach to lifting IFS with struts that they did with torsion bar IFS knuckle kits. As a result, most of the main components are the same. This Superlift system for '05 and newer F-150s consists of replacement knuckles, lower control arm crossmembers, differential and sway-bar drop brackets, and a few other odds and ends. The main difference is the absence of torsion bar drop brackets and the addition of strut towers (the components in the upper right and left hand of this photo). With this particular kit there is very little change to front track width, so CV axle spacers are not necessary. (Courtesy Superlift)

most of the vehicles that utilize an IFS strut arrangement also use a heavy-duty rack-and-pinion steering system. Nothing new to the automotive world, rack-and-pinion is fairly new for trucks. Though certainly adequate, rack-and-pinion does create challenges for lift manufacturers.

Lift System Basics

Lifting a strut IFS system starts out the same as traditional IFS in that many key components are relocated downward, thus gaining lift. The lower control arms are relocated downward via plate-steel crossmembers, while the front differential housing is relocated the same amount to keep CV-axle operating angles in check. In virtually all cases, the upper control arms remain in their factory locations and a heavy-duty replacement steering knuckle is supplied to span the distance between the control arms. These replacement knuckles also incorporate a raised tie-rod mounting boss, since relocating the bulky rack-and-pinion steering system used on most of these trucks is impractical. Various other items—including sway-bar links, skid plates, and lateral reinforcement bars—round out these lift systems, providing a fairly simple package common among all reputable suspension manufacturers. In

Paying close attention to how lift manufacturers address the struts with a lift system provides clues as to how the final-product ride and handling will be. With this spacer installed atop the strut, factory ride characteristics, both good and bad, are retained. Other forms of modification may change the valving and spring rate to compensate for heavier tires and wheels, and to improve off-road performance.

Traction control is nothing new, but the latest generation of vehicles benefit (or suffer depending on your opinion) from highly sensitive traction control systems. Called by a variety of different acronyms, traction control is here to stay and is only going to get more advanced and therefore more sensitive. The problem is that the same traction control features that can help on the street can be dangerous in an off-road environment. For now, most vehicles have a button (with often vague labels like electronic stability program, or ESP) somewhere on the dash like this one that allows the user to at least partially disable the system. The question is how much longer will these override buttons be here?

most cases, there is little if any change in the amount of suspension travel, so do not expect any gains beyond what the factory configuration allows.

The major departure among the different lift systems offered has to do with how the struts are addressed. The struts attach to a "tower" on the side of the frame and the lower control arms. With most basic systems, a spacer or "hat" attaches to the top of the strut and spaces down the factory assembly. This technique ensures factory ride-quality is maintained. Some systems place a spacer *under* the top of the strut, which preloads the coil

spring. Still other methods replace some or all of the factory strut components. How the suspension lift addresses the strut is the real key to how the truck rides and performs once the lift is complete, so each method is examined in detail.

One last item that is important to remember with this suspension design is that virtually all of them sit very low in the front as compared to the rear, often more than 2 inches lower. While this eases vehicle entry and egress, it also means that the net amount of lift often ends up looking like less than that of the same model from the previous generation. In other words, a 2005 F-150 may sit a little lower than a 1999 F-150 with the same amount of lift installed. This also has an impact on rear lift: since almost all aftermarket suspension manufacturers end up leveling the stance of the truck, do not be surprised if a 6-inch lift comes with only 4-inch lift blocks for the rear.

Traction Control: New Challenges

Once upon a time, the main concern as far as electronic compensation for larger tires was speedometer correction, if the vehicle owner even cared about that. Not so anymore. Just as vehicle suspensions get more advanced, so do vehicle electronics. In light of some safety concerns that have come to the surface in recent years, vehicle manufacturers are gradually implementing highly advanced traction-control systems to trucks and SUVs. These systems get their information from multiple sensors and systems on the vehicle and attempt to save the driver from him- or herself if the system senses a loss of traction, or if vehicle roll and yaw

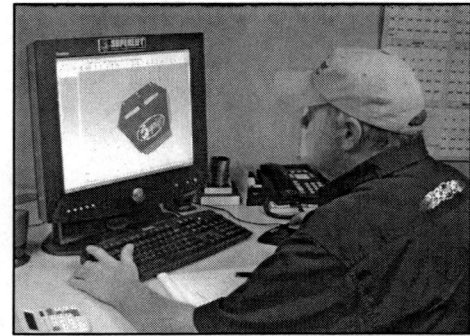

With traction control systems and more advanced designs in general, it's more of a challenge than ever for lift manufacturers to overcome the various hurdles placed in the path of designing safe and well-performing lift systems. With these new challenges, it is more important than ever for aftermarket manufacturers to work closely with OE designers. Those companies that can't or won't adapt to new technology will find themselves way behind the curve for new product.

exceed established limits. When engaged, these systems may apply the brakes to one or more wheels, reduce engine throttle, illuminate a warning light on the dash, or any combination of these things to prevent a loss of vehicle control. These systems are highly advanced and very sensitive, having been designed to step in only during the split-second window in which certain maneuvers can avoid a crash. Unfortunately, some of these systems are so sensitive that there have been scattered reports of premature system engagement or system malfunction lamps illuminating once a lift has been installed. Though it's pretty early to tell for sure, the issue seems to be either the larger tire and wheel package or minute changes in the steering ratio built into the design of certain replacement knuckles, if not both. All of the lift manufacturers are addressing these traction control issues head-on, with a select few working directly with the OE manufacturers. With some platforms (including the new Grand Cherokee and Commander), the fix is a little black box that recalibrates the system, while others are now designing their steering knuckles around the traction-control system's tight parameters.

Though traction-control problems should only be a concern for vehicles produced from about 2005 on, be sure any potential issues have been addressed when lifting a brand-new vehicle. In some cases a calibration module is needed, while others may require a quick trip to the dealer to recalibrate the vehicle computer so it knows that bigger tires have been installed. Asking a lift manufacturer directly (not the manufacturer's dealer) is always the best way to get straight answers on brand-new products.

Knuckles and Subframes

Knuckle design has not changed much despite these new advances in technology, and neither has lift method. Because moving a rack-and-pinion steering system is impractical, knuckle systems are the only choice on late-model vehicles. As a result, lift systems in the 4- to 6-inch range typically keep the upper control arm in its factory location and relocate the lower control arms and differential downward via bracketry. A replacement knuckle has a longer neck to span the increased distance between the control arms in addition to a raised tie-rod boss that provides the needed steering correction. In fact, these new strut systems are so similar to the torsion-bar systems they replaced that sometimes the lift

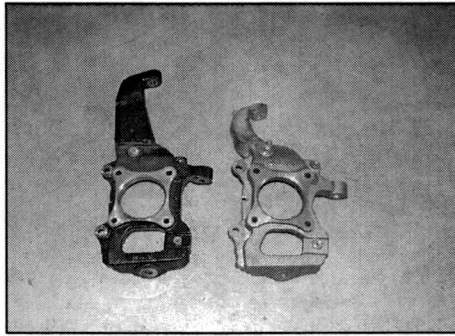

Lift knuckles on strut suspension designs are not much different from the knuckles found on torsion bar trucks. When comparing the lift knuckle (left) with the factory knuckle (right) on an F-150, the longer neck and raised tie rod mounting point are evident.

knuckle from the previous generation can be re-used. The major issue is keeping all steering ratios as close to factory as possible, as significant changes here set off the traction-control system.

As discussed in the previous chapter, a complete subframe is the mark of an exceptionally strong lift system over simple individual cross-members. A subframe locks the various frame-attachment points together and greatly reduces the chance of anything shifting or breaking when used hard. A full front skid-plate made out of heavy-gauge material helps protect the fragile front differential, which in most cases is constructed of aluminum.

Strut Matters

The whole key to the late-model IFS puzzle is how the struts (or coil-over shocks) are addressed with a lift system. Struts themselves are costly to build, and aftermarket suppliers are not as readily available as a maker of conventional shock absorbers and coil springs. In addition, the struts themselves are the central part of the suspension and are placed under a tremendous amount of pressure, so going the cheap route is not an option. Because of these factors, it is

Full skid plates like this one on an F-150 may come at a premium over base lift systems, but this extra protection can more than pay for itself when you consider the possibility of buying just one front differential housing due to trail damage. In addition to protection, the skid plate ties the two lift-crossmembers together providing additional rigidity. This Superlift skid plate has also been contoured to maximize ground clearance, which is a nice touch.

STRUT ANATOMY

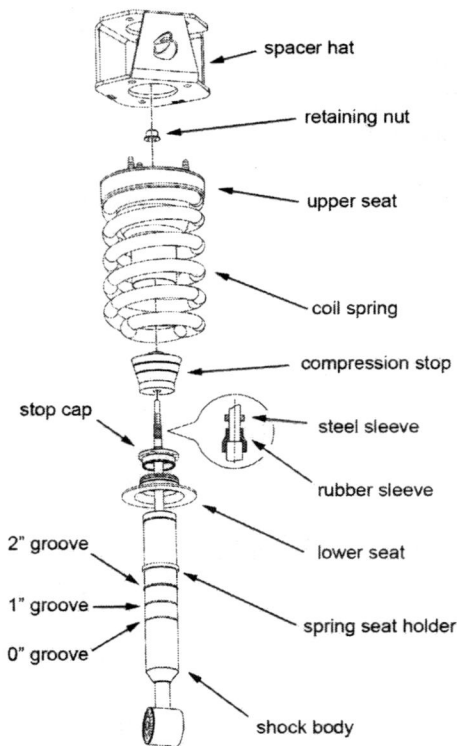

An exploded view of a typical strut.

important to understand the different ways in which a lift system will modify the struts, and equally as important, how the alterations impact performance. The three basic alterations involve strut spacers, strut preload spacers, or complete replacement pieces.

Strut Anatomy

For all of the things a strut impacts, its anatomy is pretty simple. A high-pressure monotube gas shock serves as the main body. Over this shock is a coil spring held by upper and lower retainers. Typically there is also a compression stop that rides on the stem end of the shock as well. The coil spring itself typically has a great deal of preload even when the shock is at full extension, so a special tool called a strut compressor is

One important thing to remember when working with a strut-equipped truck is that any kind of work to the strut itself requires a special tool called a strut compressor. This fixture holds spring pressure that can gradually be released once the retaining nut is removed. Remember that struts have a tremendous amount of energy stored in them and any attempt to disassemble one without the proper tools can cause serious injury or worse. Even if you get lucky and tear one apart without a compressor, it will be impossible to put it back together.

needed to take a strut apart. *Do not attempt to disassemble a strut without a strut compressor or serious injury will result.* With some systems the factory struts are left alone, while others require partial or total disassembly of the factory components, so be sure

The three studs visible on top of the strut are what normally attach to the frame. A strut spacer attaches to these studs and has a matching stud pattern at the top to attach to the frame. Therefore, the strut's attachment point has simply been spaced down to provide the required amount of lift.

Spacers are the only modification that does not require disassembling the struts themselves.

to factor in the extra time and expense of taking the struts to a qualified professional if you plan on doing the installation at home.

Spacers and Hats

With most basic lift systems, the factory strut is retained and a plate-steel or machined spacer is attached to the top. This in turn spaces down the strut the necessary distance to work at a given lift height. With this method you can be sure that factory ride-quality is retained because the strut itself has not been altered. This method also does not require any strut disassembly, as the spacer usually attaches to three or four studs on the top of the assembly. Certainly the

Preload spacers like the one installed on this strut for a Tacoma are positioned underneath the strut's top mounting plate. This places additional preload on the coil spring, thus gaining lift. Preload spacers are a good way to recover lost ride height due to heavier accessories, but they also reduce the amount of extension travel available. These are equivalent to cranking up the preload on a set of torsion bars: not a bad idea, but proceed with caution.

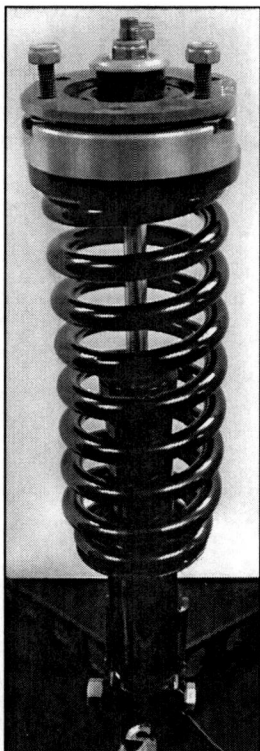

most economical method, strut spacers are a good choice for anyone who wants to maintain the vehicle's existing ride-quality. However, these spacers offer no adjustability and don't address the "spongy" feel that some late-model truck owners complain about. These spacers are typically seen on 2- to 5-inch lift systems.

Preload Spacers and Rings

Not to be confused with a conventional spacer that installs above the strut assembly, a preload spacer installs within the strut itself and (as the name indicates) places additional preload on the coil spring. Good for nominal amounts of lift and to offset a heavier tire-and-wheel package,

these should be used with caution. Just like cranking up the preload of a torsion bar, preload spacers bump up the spring rate but also reduce the amount of extension travel available (which is controlled by the shock). Any spacers that increase ride height by more than about 1½ inches will have very little extension travel left, which can cause the strut to constantly "top out." Not only does this severely impact ride-quality, it eventually destroys the (very expensive) strut. Preload rings require strut disassembly and install either on top of the coil or underneath the lower spring seat. Some manufacturers combine a preload ring with a conventional spacer to both gain lift height and firm up the ride-quality for a more solid feel on the street.

Replacement Shocks and Springs

Though a more expensive option than a simple spacer, replacement shocks, springs, or complete strut assemblies are the best way to increase the off-road prowess of the suspension. They also happen to look pretty cool. Which components are replaced depends on the model, the lift manufacturer, and the "level" or "stage" of the lift system. In the case of a replacement shock, all factory components are transferred over to the new piece, which usually has a longer body (for lift) and valving better suited for off-road use and larger tires. With a replacement spring, some lift is gained, but the new piece has a better rate to work with the lift system. Complete replacement struts, while the most expensive option, are also the way to go if extended off-road forays are in the truck's future. Designed to survive in punishing environments, some of them are derived directly from race

coil-over technology with many of the same attributes, such as being adjustable and rebuildable.

Shorter Lifts

Up until this point the information has been focused largely on 4- to 6-inch kits, but what about the smaller systems? Since strut-equipped trucks sit very nose-heavy anyway, what about just gaining some mild altitude and some performance at the same time? The fact of the matter is that strut designs are very receptive to mild lift systems, provided that they address the needed point—namely alignment and CV-axle specs. It's also important to understand that not all mild lift methods are created equal. Generally speaking, spacer lifts above the 2-inch level yield a streetable truck with no performance gain, while lifts that address shock length, valving, and spring rate better serve the needs of off-road enthusiasts.

Longer shocks are needed for the reasons stated earlier: you're not really gaining height by preloading or replacing the spring, all you're really doing is robbing extension travel. A replacement shock restores the extension travel to factory specs, but using a factory coil spring might be a problem for those looking to improve the ability of the truck to perform in high-speed environments. The best option for those looking to gain *both* suspension travel *and* off-road-ability should strongly consider a shock *and* a coil spring replacement. However, 2½ inches of lift is generally riding the ragged edge of factory geometry without the associated drop brackets to alleviate other critical suspension components.

Tacoma Solutions

Toyotas have always had a loyal following and a well-deserved reputation for longevity. They even have a solid foundation in the off-road world thanks to factory-offered lockers and overbuilt drivetrain components. The one area they could use some help in, however, is the ground-clearance department because no amount of traction is going to overcome getting hung on the frame with the tires dangling in the air. Fortunately, Skyjacker has the perfect solution for Tacoma and Tundra owners wanting a mild boost to clear moderately bigger tires. Its Platinum Series coil-overs replace the factory strut and its mediocre performance. These are true race-inspired shocks that offer adjustability to compensate for things like a heavier front bumper with a winch. This upgrade adds up to 2½ inches of lift and a better spring rate as well as valving for when the pavement ends. Best of all, the whole procedure can be done with common hand tools and doesn't take much time.

The procedure that follows can be done on any '96–2004 Tacoma or Tundra, and a very similar upgrade is available for the '05–current Toyota pickup trucks. A racier look, more ground clearance, and a better ride–who could ask for anything more?

The heart of Skyjacker's 2½-inch lift system is its Platinum Series coil-over front shocks. If these units look like race shocks, that's because they are derived straight from racing technology. Skyjacker paid close attention to proper spring rate and supplies a new coil spring as well. By replacing both the coil and shock rather than just one or the other, you can be sure this system was designed with off-road use in mind. Oh yeah, this stuff looks pretty cool, too!

A mild look but major gains, this Tacoma is ready for the trails!

1. The coil-over assembly ships unassembled, so the first order of business is getting the shocks put together. This requires a strut compressor, which is a special tool not available in the average shade tree mechanic's garage and not even ordinary repair shops. It also takes some training to operate one properly, so this portion of the assembly is not for the do-it-yourselfer. Many four-wheel-drive shops have one, or any shop that does a lot of work with struts. The fee is usually nominal, as it doesn't take much time to do the work. The coil-over should be assembled as shown. (Larry Conville/Skyjacker)

Tacoma Solutions *continued*

2. The shock body has a long section of external threads, and two spanners serve as the upper spring seat to provide the necessary ride height adjustments. The instructions recommend backing off the upper spanner all the way to the top of the threads and then adjusting the lower spanner until the top of it is 1⅝ inches away from the top of the upper spanner. This provides a baseline ride height setting. It is important for both coil-overs to be adjusted the same amount.

4. With the new shocks ready to go, it's time to take the truck apart. Secure the frame on jack stands so the front suspension is at full extension travel, and remove the front tires. Detach the lower end of the factory strut from the lower control arm using a 19-mm wrench. Next, remove the three nuts on the studs securing the upper end of the strut to the frame using a 14-mm wrench. The strut can then be removed and discarded.

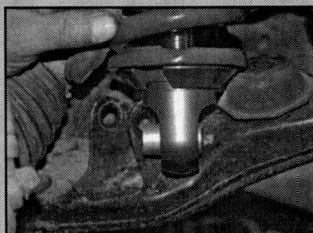

6. Pry downward on the lower control arm enough to attach the lower end of the shock to the control arm at the factory mounting location. Be sure to position the supplied spacers as shown, with the thicker of the two pointing forward toward the CV shaft. Secure the shock using the factory hardware.

3. Attach the upper mounting plate to the eye on the main body of the shock. An Allen bolt is supplied for this purpose, but be sure to install the two short misalignment spacers on either side of the heim end prior to bolting on the plate. Be sure this bolt is tightened, as it will not be accessible later.

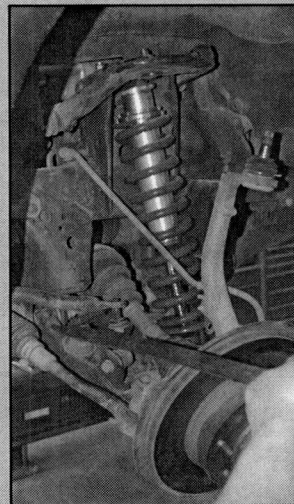

5. Unbolt the sway-bar links from the lower control arm. Remove the nut on the upper ball joint and, using a special puller tool or a pickle fork, separate the upper ball joint from the knuckle. Next, attach the upper end of the shock to the frame using the supplied fine-thread hardware. Be sure that the shock is indexed as shown, with the upper bolt on the shock parallel with the frame.

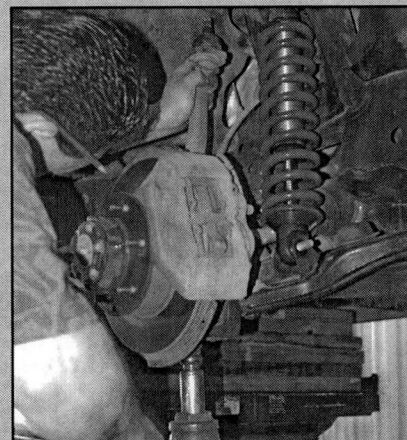

7. Place a bottle jack under the lower control arm and raise it enough to re-attach the upper ball joint to the knuckle. Torque the ball joint nut to factory specifications and re-install the sway-bar links.

Tacoma Solutions *continued*

8. Skyjacker supplies a spanner tool designed for use with a ⅜-inch ratchet. This spanner tool enables ride height adjustments to be made. After locking the upper spanner to the lower, reinstall the tires and lower the vehicle to the floor. Take some ride height adjustments, and if the front needs to go up, tighten the spanners. If it needs to go down, loosen the spanners. Only make spanner adjustments with the suspension at full extension, and always lock the upper spanner before loading the suspension. Also be sure both shocks are adjusted the same amount. Repeat this as necessary until the desired ride height is achieved.

10. Place a C-clamp on either side of the spring tie bolt as shown. Once the spring pack is firmly clamped together, use locking pliers and the appropriate-size wrench to remove the tie bolt. Very carefully loosen the C-clamps to separate the spring pack. Use caution here because if the C-clamps slip or fail, the spring pack will come apart with explosive force!

12. The last step in the process is installing the new shock absorbers after securing the brake hose to the axle. Before lowering the vehicle to the ground, double-check all fasteners for proper torque and all components for proper clearance.

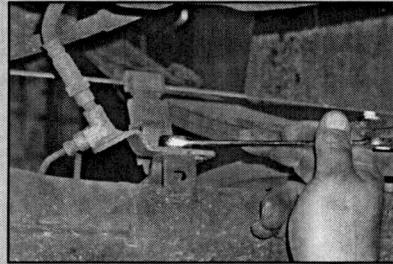

9. For rear lift, first unbolt the rear shock absorbers from the axle and then unbolt the brake hose from its attachment point on the axle. This enables the axle to be lowered enough to facilitate installing the add-a-leafs.

11. The supplied add-a-leaf installs directly above the flat overload leaf. Place it in the spring pack with the long end toward the rear, and then use the C-clamps to compress the spring pack. Install the new supplied tie bolt and tighten it to secure the pack. Do not use the tie bolt to compress the spring pack! Once installed, trim off the excess length and remove the C-clamps. Position the bump stop bracket as shown and then re-install the U-bolts.

The finished product is subtle, but the performance improvement is not. The Tacoma went from acting like a pogo stick on washboard roads to soaking up surface irregularities with ease. New tires and wheels round out the upgrades.

REAR SUSPENSION

Up until this point, the entire focus of this book has been the front suspension, but of course that is only half of the overall equation! The best-performing, well-designed front suspension does nothing without an equally adequate system in the rear. While many people approach the back of their rig as an afterthought, it's an important part of the overall package.

Thankfully, rear suspension is much less complicated than the front and most trucks haven't seen any major changes in decades. There is no steering to worry about, and solid axles rule the roost, so things are kept pretty simple. There are only two major variants: leaf and coil spring; Independent Rear Suspension (IRS) has been kept largely at bay, though it has started popping up in a few SUVs. However, do not make the mistake of thinking that lifting a rear suspension is not without its challenges. This chapter closely examines both rear suspension types, identifies the advantages and shortcomings of each, and explains how they are modified for better off-road performance.

Even though it's a 1-ton truck, the rear suspension of this Dodge Ram flexes surprisingly well on an RTI (Ramp Travel Index) ramp. Thanks in part to its simplicity, not a lot of modifications are necessary to make the rear suspension perform well.

Leaf Springs: The Gold Standard

Leaf springs and four-wheel-drive trucks have been together since the first 4x4s rolled off the assembly line, and that partnership continues today for the rear of all pickup trucks and the majority of SUVs. There is simply no better design that can adapt and perform well in the multiple roles a pickup serves in: whether

The rear suspension on a 4x4 has not changed much from what is shown here: leaf springs and a solid axle. The only variations are shock position (the rear of this Chevy has the shocks staggered to combat axle wrap), whether or not a sway bar is present, and how the rear brake hose is positioned. This simple configuration is easy to modify and lift height is only limited by driveline angle.

Perhaps the greatest impact on the ride quality of the rear suspension is the designed weight rating of the truck. A ½-ton truck is supposed to be able to haul just that, while a 1-ton is designed to carry much more. Since rear spring rate is determined by the capacity of the pickup, the heavier truck is going to have a firmer ride. This F-450 is a step above a 1-ton, so it is going to ride noticeably stiffer than an F-150. Though this may be obvious to many, don't expect a lift system to smooth the ride on a heavy-duty truck.

it's hauling, towing, or just commuting, leaf springs can do it all. However, the one area in which they can be troublesome is in ride quality; a quick ride in an empty ¾-ton truck leaves most people searching for a few bags of cement to throw in the back to force some give in the rear suspension. The fact of the matter is that it is impossible to create a leaf spring that rides comfortably on the street and support the weight of hauling and towing duties truck owners demand. A soft-riding spring easily sags under a load, while a spring designed around load-carrying capacity loosens fillings. It is a delicate balance for any spring designer to hit an acceptable compromise between ride and load capacity. Often an overload spring is utilized to increase the rate of the springs once a load is in the bed without impacting unloaded ride

quality. However, the important thing to remember here it that spring rates err to one side or the other depending on the weight rating of the truck. This is why a ½-ton truck always rides better than its heavier ¾- or 1-ton brethren. Though it seems obvious, choosing the proper truck for individual needs is all too often forgotten as people complain about the harsh ride of their new 1-ton. Generally speaking, stick with a ½-ton if ride quality is a concern, or live with the firmer ride of a ¾- or 1-ton if that's what's needed to safely haul heavier loads. Along with this, it's important to remember that a lift system is designed around the same weight rating as the truck, so ride quality is not usually going to improve with a lift system.

There are three ways to lift the rear of a leaf-sprung truck: a block, a

block and add-a-leaf combination, or replacement leaf springs. Each one has its pros and cons.

Lift Blocks

Universally frowned upon in off-road enthusiast magazines, but contrary to most opinion-leaders, lift blocks are often the best choice for a truck. Lift blocks install in between the bottom of the springs and the axle pads, essentially spacing the springs farther away from the axle to achieve lift. They have a hole on one side to accept the center pin of the spring and a pin on the other to engage the hole in the axle pad, while longer U-bolts clamp the blocks in place. They are the most economical lift method by far, are usually included in "basic" or "standard" lift systems, and thousands of them are sold every year for use in the back of pickups and SUVs. Even

A lift block (the rusty part) is nothing more than a spacer positioned between the factory leaf springs and the axle. A typical block kit also includes new U-bolts, but the factory "saddle" or retaining plate is retained. Though illegal in some states, blocks do allow the factory springs (and their corresponding ride quality) to be retained. Because of this, not many replacement spring options are available for 1-ton trucks. Lift blocks come in a variety of heights but should never exceed 5 inches.

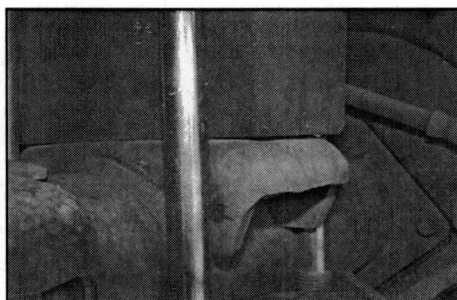

Many trucks come with short lift blocks straight from the factory. Some incorporate a compression stop, while others (like this one) are specially designed to conform to the axle spring pad. Unfortunately, in this situation using both the factory block and a lift block is needed in order for the blocks to be adequately seated on the spring pads. Stacking blocks, however, is never a good idea and a recipe for failure if this truck ever gets used hard in the rough. This situation is the only time stacking blocks is even mildly acceptable; never, ever stack two aftermarket blocks.

Here is the reason factory blocks must sometimes be retained. Without it, this lift block is barely seated on the spring pad. Without proper pad-to-block engagement, there's a risk of the block breaking or "rolling" off of the pad. This can lead to major damage at best, or at worst, an accident.

so, the off-road community reviles blocks, and in a couple of states they are even illegal. What gives?

The Pros: Since the factory leaf springs are retained when lift blocks are used, so too are the factory ride quality *and* load-carrying capacity. These are things that neither of the other two methods can claim, and since most trucks frequently haul or tow in addition to serving as commuters, these are both very important attributes that should be retained. As mentioned earlier, lift blocks are also the most inexpensive lift option.

The Cons: Lift blocks amplify the amount of leverage placed on the rear springs. On vehicles with lots of torque on tap (such as diesels), this increased leverage can cause problems with axle wrap and driveline vibration. It can often take quite a bit of tuning to get rid of these prob-

lems, and sometimes a set of traction bars is the only solution (more on these later). This leverage is also a big problem in the off-road camp. Off-roading introduces forces that are never experienced on the street, and the greater leverage on the suspension caused by using blocks can make the blocks "roll" out from their perches, usually with disastrous results. The taller the block, the greater the likelihood that it will fail. The design of the spring perches on some trucks is not very compatible with aftermarket lift blocks, forcing the lift block to be "stacked" on top of a factory block. Stacking blocks is never a good idea and is usually illegal. Speaking of illegal, a few states (Pennsylvania is one) have outlawed the use of rear blocks; so check local

laws before purchasing a lift system (outlawing the use of blocks seems a tad silly since many trucks come right from the factory with rear blocks installed, some of them more than 3 inches thick!). Aluminum lift blocks should be avoided, and never, ever install lift blocks on the front of a leaf-sprung truck.

Blocks and Add-a-Leafs

For mild amounts of lift, add-a-leafs are a viable option by themselves, but with lifts above about 3 inches, it is usually necessary to combine an add-a-leaf with a short block. As the name implies, an add-a-leaf is a single leaf that is installed with the existing spring pack to provide additional arch and rate to the factory spring. There are two basic kinds of

Add-a-leafs come in a variety of shapes and sizes to accommodate different spring lengths and arches.

Installing an add-a-leaf requires disassembling the spring pack. The leaf should be positioned directly under the factory leaf that is a little longer than the leaf being added. With long add-a-leafs this is usually directly under the main leaf, while short ones usually attach to the bottom of the pack. When purchasing an add-a-leaf kit, be sure that new spring tie bolts are included. (Courtesy Larry Conville)

add-a-leafs: short and long. A short add-a-leaf is usually positioned toward the bottom of a spring pack (often just above the overload leaf) and can add quite a bit of rate to the spring. A long add-a-leaf usually installs just below the main leaf in the pack to increase its arch. Generally speaking, longer add-a-leafs have less impact on ride quality and load-carrying capacity than a short add-a-leaf, but both styles create a noticeably firmer ride.

Add-a-leafs are a great choice when additional load-carrying capacity is needed or when the factory springs are marginal in terms of maintaining ride height. Combining them with a short lift block is a good strategy for gaining the necessary amount of lift to match the front with less block height than using blocks by themselves. Add-a-leafs are also inexpensive and the best choice for someone on a budget who wants to minimize or eliminate the use of blocks completely. However, ride quality reductions should be expected.

Replacement Springs

The most expensive option, rear springs are the best choice for a vehicle that is going to spend a lot of time off the pavement. Replacement springs allow the rear suspension to perform as good as—and often better than—it did in its factory configuration. Axle wrap and driveline vibration are also rarely a problem with rear springs, as the rear axle has no more leverage than it did with the original springs. Simply put, off-road vehicles should have rear lift springs. Rear springs are also a good choice for a vehicle with worn-out factory springs, and are the only choice for

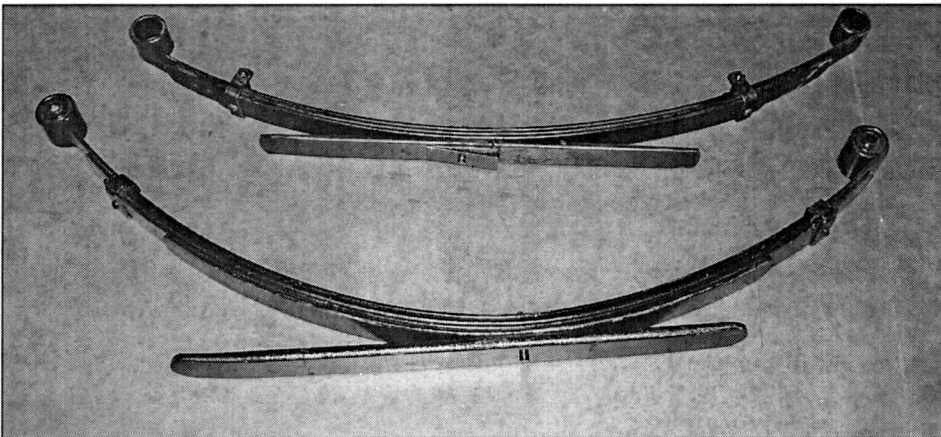

Replacement springs are always the best option for off-road use and are required on any vehicle where the axle is positioned above the springs (such as on Jeep CJs and YJs). The only downside is that more arch means reduced ride quality because the leafs within the pack must slide farther against one another; more than in a flatter pack. Therefore, you can expect a more compliant ride with the spring at the top than the one on the bottom.

The thick flat bottom leaf is known as the overload, and the greater arch of a lift spring means that the spring pack must travel quite a ways before the overload engages. This results in more squat when a load is placed in the bed. A good way to combine load-carrying capacity and a lift spring is to install a set of airbags that can be inflated for more load-carrying capacity when the vehicle is loaded and then deflated for a better ride when unloaded.

Just like the front, longer shocks are almost always required for the rear of the truck when installing a lift system because the axle is moved farther away from the frame. Shocks come in a variety of styles and should be properly valved to work with the rear suspension. In this case, monotube gas shocks were chosen, but any quality hydraulic shock will work well with rear leaf springs.

those applications in which the springs attach to the bottom of the axle rather than the top (such as Jeep CJs and YJs).

The flip side of rear springs is that ride quality and load-carrying capacity both suffer. The increased arch of a lift spring means it must travel farther before a bottom-mounted overload can be engaged, and lift springs are usually incompatible with a top-mounted overload. As a result, hauling at or near the truck's capacity can result in headlights pointed at the moon. As far as ride quality, more arch causes more friction between the individual leafs of a spring, which has the effect of feeling like more spring rate is there without actually having any.

The Bottom Line on Lift Method

The correct method of rear lift must be dependent upon the intended use of the truck. If it's primarily street use and frequent hauling or

towing in the plans, stick with blocks or a block and add-a-leaf if ride quality is less of a concern than weight capacity. Rear springs, though expensive, are a must for any dedicated off-roader.

Shocks and Other Necessities

Regardless of lift method, anything over about 2 inches of lift is going to need replacement shocks. As mentioned in Chapter 3, some questionable mail-order advertisers do not include shocks in an advertised price for older vehicles, so be sure shocks are included. As for the type of shock, there are many different manufacturers with many different kinds of shocks, and it really boils down to which advertising claims to believe. A good tactic is to ask around and then get the nicest shocks your budget allows.

It seems strange that many lift manufacturers offer systems that are very complete for the front suspen-

Top-mounted overload springs are more common on heavier trucks. With this type of overload, the leaf mounted above the main pack engages tabs on the frame as the suspension is compressed. Unfortunately, the greater arch of a lift spring usually interferes with this type of overload and it must be removed. This can severely reduce the load-carrying capacity of the truck. Identifying which type of overload the truck has is also important when installing lift blocks, as top-mounted overloads usually require longer U-bolts than conventional overload leafs.

It's often the little things that make the difference between an average lift kit and a well-designed one. Lift companies often forget the rear compression stops, but they serve an important role in keeping the tires away from sheet metal. Often a simple spacer is employed to lower the factory stop. If your lift kit didn't address the rear stops, it's usually pretty easy to fabricate your own.

sion and then completely drop the ball out back. Perhaps wooed into laziness by the rear's relative simplicity, there are still many items that must be addressed. In most cases the rear brake hose must be relocated or replaced entirely, yet many of them neglect to say anything about it. The same goes for the rear axle breather. Virtually all of them are guilty of neglecting the rear compression stops on at least one application, which means the new larger tires are free to mangle sheet metal the first time the vehicle is used hard. The emergency brake cables often require re-routing or relocating. A few vehicles, mostly Toyotas, have a proportioning valve for the rear brakes that adjusts pressure to the rear axle as more load is placed in the bed. This valve is hooked to a rod attached to the rear axle, and it too must be corrected when more ride height is added or braking will suffer. Making sure all of these items are addressed leads to a

more satisfactory finished product.

The one wild card to a rear leaf-spring suspension is the presence of a sway bar. Not all trucks are equipped with a rear sway bar; it usually depends on which option box was checked when the vehicle was built. Certain towing packages and "camper specials" may include a rear sway bar, but because they are not normally standard equipment, most lift manufacturers do not address them. Depending on the intended use of the vehicle, some people simply remove them, but lifting the truck will minimize or negate their effectiveness if left unattended. Cutting and extending the factory sway bar links or adapting a set of extended links is really the only option if a rear sway bar is retained.

Coil Springs

Aside from some early two-wheel-drive models, rear coil springs are new to the rear suspension of pickup trucks and SUVs. Usually the result of the ever-important quest for ride quality, rear-mounted coil springs represent a boon for both off-road performance and ride quality if properly executed.

As has been discussed previously and unlike their leaf spring counterparts, coil springs have no structural properties, so it is up to other components in terms of links and bars to properly locate the axle under the vehicle. Though this does add complexity, a coil-spring suspension is surprisingly lenient to lift kits as long as the lift in question maintains certain critical geometry. Most evident in late-model Jeeps and SUVs, coils (and their air-bag variants) offer both excellent ride characteristics and a supple, trail-conforming platform to

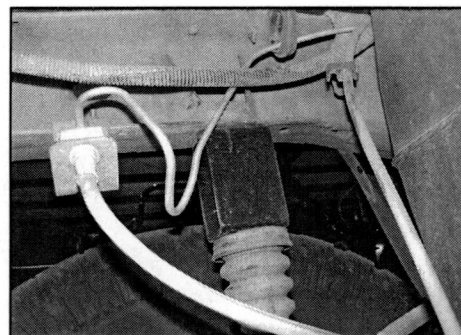

Making sure the rear brake hose has an adequate amount of slack is critical when lifting any vehicle, and once again, this is where some lift companies drop the ball. Depending on the design of the rear brake hose, sometimes a simple bracket is necessary to space down the attachment point of the factory hose. In this case, the factory bracket was simply moved from its original location on top of the frame to the bottom. A replacement brake hose should be seriously considered for those intending to go off road.

soak up the whoops and provide excellent slow-speed off-road handling traits.

Lift System Basics
In virtually all cases, a four-link system supplies the necessary axle location along with a track or panhard bar to control side-to-side movement. Like lift kits in general, the amount of modification needed to work properly depends largely on the amount of lift in question. At lower heights, say 2 to 4 inches, very little is needed beyond taller coils and shocks. The deviance from the basic mods happens at lift heights above 4 inches in general and is further complicated by the factory geometry and wheelbase. Generally speaking, short-wheelbase vehicles like Wranglers need more correction

Rear coil spring suspension is quickly making a name for itself in the off-road world thanks in no small part to the Jeep TJ and JK (the rear of a TJ is shown here). They offer a better ride and also superior suspension travel, though these benefits come at the expense of a greater tendency to exhibit body roll.

Axle Wrap

Combating axle wrap has always been a problem with a leaf-spring suspension, and over the years there have been many different devices that have come and gone designed to reduce or eliminate it. Short of improper overall geometry, coil-spring-suspension axle wrap is the result of worn bushings or other components.

The Problem

Axle wrap occurs when torque from the engine (via the rear drive-shaft) forces the rear axle pinion to climb (or dive in the case of a front axle). If enough torque is applied, this force causes the rear springs to bend from their natural arch into an "S" shape. The springs introduce resistance to this deviation from its natural shape and add pressure in the opposite direction (back to its natural arch). If torque continues to be applied from the engine and the pinion continues to climb, the spring

sooner, while longer-wheelbase vehicles like Tahoes and Expeditions are more accommodating and need very little in the way of modifications. Regardless, the key to all of them is maintaining proper geometry if the goal is a good balance of off- and on-road driving characteristics (not necessarily in that order).

Lifting a coil-spring rear suspension system is deceptively complex because not only is ride height a consideration, but so too is overall geometry, wheelbase change, and driveline angle. Lift manufacturers approach this by a variety of different means, from just taller coils to relocation brackets for the factory links to a pair or complete replacement links to keep the factory geometry intact. Regardless, the factory settings are the ultimate goal. Therefore, this is one of the few cases in which it's important *not* to compare apples to apples. As long as the nec-

essary things happen, be it relocation brackets or replacement links with taller coils, it's all good. What is important to remember here is that *something* is included.

With a rear four-link system, upper and lower links run between the axle and the frame while a track bar handles side-to-side duties.

Addressing the track bar is an important consideration on the rear suspension, as leaving it alone creates all sorts of negative handling traits. In this case (a Jeep JK), a simple bracket raises the attachment point on the axle. Replacing the stock bar with an aftermarket adjustable one is another viable option.

This can be the end result of axle wrap if you're not careful. The shock loads from axle wrap caused a rear axleshaft to snap in two, but unfortunately it broke and exploded inside the differential carrier and destroyed it as well. To make matters worse, depending on the design of the rear end, a broken axleshaft can cause the tire to separate from the vehicle. Essentially you're looking at $1,000 worth of parts and labor to fix this axle. Ring and pinions, driveshafts, and transmissions and transfer cases are all susceptible to damage from axle wrap.

energy eventually overcomes the traction of the tires on the ground and the springs jump back to their normal arch. This forces the axle back downward, usually in a violent release. This can happen several times within seconds and the end result is a "pogo stick" effect in which the tires catch and release traction as the springs wind up and release their energy. This causes a "wham-wham-wham-wham!" sensation from the drivers seat until one eases up on the throttle, the tires break traction completely, or the torque of the engine and the torque in the springs equalize and gradually dissipate wind-up as the vehicle accelerates.

It is important to remember that many trucks have an axle-wrap problem when in stock form. Simply side step the clutch or hammer the throt-tle from a stop and most trucks reveal some form of axle wrap. Unfortunately, the increased lever-age of larger tires, lift blocks, and increased driveline angles usually aggravate an existing problem. In many cases, judicious driving is the best defense. But the real hindrance of axle wrap for off-roaders is two-fold: traction can be severely com-promised (remember that traction is the ultimate goal), and severe axle wrap can break parts because it causes extreme shock loads to the drivetrain. In nearly all cases, drive-train parts are expensive.

The Solution

Unfortunately, there is no easy cure for axle wrap. Because many vehicles exhibit axle wrap stock and a lift and larger tires only makes it worse, often you are "fixing" an existing problem. The easy answer is judicious driving and understanding the limits of the vehicle in question. The more complex answer lies in a delicate balance of driveline angle

The shorter the block, the less severe axle wrap will be. Reducing the height of the block used by either reducing lift height or installing an add-a-leaf and a shorter block can often solve axle wrap issues, or at least convert a big problem into one that is livable.

(discussed later) and equipment. Rear lift blocks place greater leverage on the rear suspension, which is why they are the most common culprits for axle wrap. Short of knowing the vehicle's limits and keeping the throt-tle light in certain situations, the best fix that does not hamper ground clearance is utilizing add-a-leafs or

Traction bars are the most effective way to eliminate axle wrap. Though off-road enthusiasts are generally not fans of these bars because they reduce ground clearance, the street guys like them because traction bars look pretty cool. With high-torque engines and large tires, a quality set of traction bars may be the only effective solution. Though it is tempting for some people to build their own (and indeed, many homegrown versions have been built that work very well), sticking with store-bought versions like this one is the best choice for those that don't have advanced mechanical skills. Note this bar has a pivot near the frame that allows the bar to twist when the suspension is loaded on one side.

replacement rear springs rather than blocks. Add-a-leafs almost always help reduce axle wrap because they increase rear-axle spring rate, but this can hamper overall driveability. Replacement lift springs almost always solve an axle-wrap problem unless a very high-torque diesel engine is involved or there's a spring-over application involved.

Assuming driveline angles are correct, the only way a block-induced problem can be solved is with a set of traction bars. These are bars that attach to the frame and a point on the axle (usually below the axle's center-line). Traction bars create a fixed point that prevents the rear axle from rolling under acceleration. While an effective solution, traction bars are a tricky proposition, as they must be designed to do their job without hampering suspension travel. An improperly designed set of bars locks down the rear suspension and allows little if any movement, which results in an incredibly harsh ride. A properly designed set of bars allows the rear suspension to move freely, but prevents the rear axle pinion from climbing under torque. The other negative is that most traction bar designs reduce ground clearance because they hang below the frame and axle. While not a big deal for street-bound trucks (and indeed, traction bars usually "look" pretty cool), this is a real problem for a dirt-bound truck. A better design for off-road trucks is a set of bars that attach above the centerline of the axle; unfortunately, few if any bolt-on traction bars are designed this way. That leaves mostly homegrown solutions and a lot of trial and error to get the geometry of the bars just right. While many ingenious and effective systems have been fabricated in garages, many more have been done that are either ineffective or limit travel. Short of finding a bolt-on system, be prepared for a deceptively difficult task if you plan on making your own.

Though axle-wrap problems are common in the truck and Jeep worlds, the solutions are often very different. The Jeep crowd is much more sensitive to ground clearance because a Jeep's short wheelbase makes a traditional bar less effective since they are too short to be really effective. Though most bolt-on lift springs do an adequate job controlling wrap-up, Jeepers usually add big

In the constant quest for lift and off-road performance, many leaf-spring Jeep owners opt for a spring-over conversion in which the axles are moved from above the springs to below. This provides about 5 inches of lift and increases ground clearance, but a spring-over is not as easy as welding a new set of perches to the axle. Stock springs are not up to the task of controlling axle wrap with a spring-over, making significant changes to the springs or some sort of traction device mandatory.

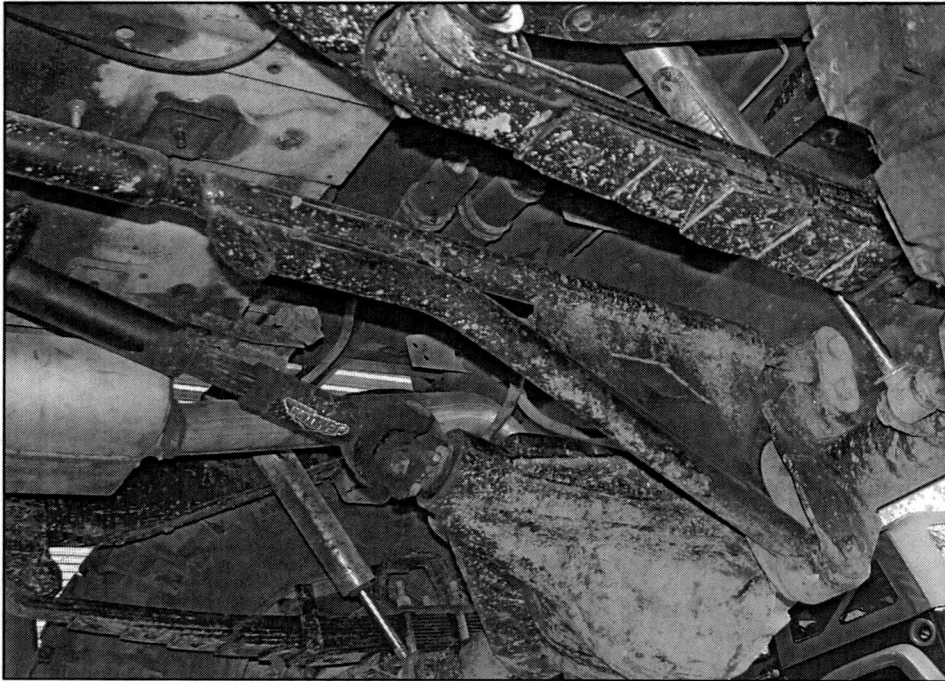

Moving to a view in front of the axle on this spring-over Jeep reveals the most common type of traction bar used on Jeeps these days. An effective bar on short-wheelbase vehicles involves attaching it above and below the centerline of the axle, otherwise the springs serve as half of the traction bar "link." This bar incorporates a threaded pivot that allows the suspension to twist and also provides adjustability. This particular bar is called a Torque Fork and is made by Superlift.

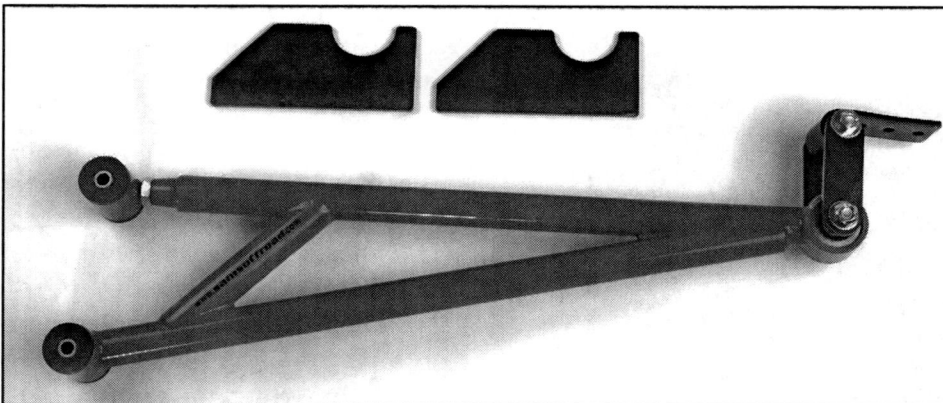

Another triangulated traction bar that performs well is manufactured by Sam's Offroad. Welded brackets on the axle supply the mounting points, and the bar's upper end attaches to a shackle mount. The other end of the shackle is attached to a crossmember added to the frame. This bar also features a Johnny Joint at the upper end to absorb twist as the suspension articulates and has a threaded upper axle mount that allows the bar to be "timed" properly with the suspension. Essentially, this bar allows the axle to move wherever it wants while preventing it from "rolling." (Courtesy Sam's Offroad)

power gains (in the form of engine swaps) and some dabble with placing the axles below the springs in a form of lift commonly referred to as a spring-over. The main issue with a spring-over is that moving the axle below the springs introduces them to leverage they were never designed to handle. Stock leaf springs are not capable of handling the greater leverage and mild lift springs are only marginally better. To make matters worse, beefing up the spring pack enough to control axle wrap usually results in a very harsh ride. To further add fuel to the fire, a traction bar that attaches to the axle at a single point still relies upon the spring as a second "lever," and soft springs usually end up bent and sagged over time.

The effective solution for a Jeep is slightly more complex than a pickup but must satisfy *all* of the requirements for the Jeep crowd: control wrap-up, minimize the impact on ground clearance, and have very little impact on the rear suspension's ability to articulate. Though several effective homegrown solutions have been created, in recent years one design has emerged as the best solution: a single "Y" shaped bar that attaches above and below the centerline of the axle and extends to a crossmember added to the frame. The bar then attaches to a shackle mounted to the crossmember and has some method that enables the mounting points to twist (usually via heim joints or hybrid bushings like Johnny Joints). This design allows the rear suspension to do anything it wants except wrap up, so the overall impact on articulation is minimal. These bars are currently available through places like Sam's Offroad and Superlift to accommodate a variety of different axles.

Short wheelbase vehicles are particularly susceptible to driveline vibration because the rear driveshaft is also short. The shorter the shaft, the more quickly angles change as lift is added. Even with factory axles, the tubing in this TJ's driveshaft is just a few inches long, with the rest made up of the slip shaft and U-joints.

Older vehicles equipped with manual locking hubs like these allow the owner to truly disconnect the front axle and driveshaft from the transfer case. Alas, it was apparently too much trouble for drivers to manually engage the front axle so vehicles these days don't have locking hubs. The by-product of this is that the front driveshaft spins all of the time and can be the vibration source. On vehicles equipped with manual or "auto-locking" hubs, it's a good idea to verify that the front driveshaft is truly free in two-wheel drive as they are known to stick every once in a while (especially the automatic versions).

Driveline Vibration

Though an independent problem, axle wrap and driveline vibration are often closely related. This is because vibration (caused by the deflection of the springs due to axle wrap) can introduce mild-to-major vibrations under certain conditions. Even if the truck does not present the violent "wind-and-release" end result of axle wrap, improper driveline angles can create a "launch shudder" in which mild-to-severe vibration can be felt when accelerating from a stop. Though usually felt at slower speeds, driveline vibration can occur at any time.

Diagnosing the source of driveline vibration can be difficult and frustrating; with the solution for one truck not always working on another even if the two are nearly identical. There are two basic kinds of vibration: constant (regardless of speed, under both acceleration and deceleration), and intermittent (exhibiting under acceleration, deceleration, or a certain speed but not all the time). As lift height increases, so too does the risk of vibration because the higher the lift, the greater the driveline angles. Most full-size trucks can handle 6 inches of lift before vibrations become a concern, while Jeeps can usually handle 4 inches with stock components. Above these thresholds, troubles with vibration are much more common, and extreme lift often requires extreme measures to keep the axles linked to the drivetrain.

Front vs. Rear

First off, don't always assume the vibration you feel is coming from the rear driveshaft. Though most older vehicles disengage the front driveshaft in two-wheel drive via locking hubs, most late-model vehicles keep the front driveshaft spinning in two-wheel drive even if it's not transferring power to the front wheels. Many others have a full-time four-wheel-drive option (such as GM's Autotrac system) that sends a certain percentage of torque to the front axle. As a result, it is best to verify which driveshaft is causing the problem. Some full-time transfer cases have a "true" two-wheel-drive function, so with those vehicles try seeing if the vibration can still be felt in two-wheel drive. Failing that, the most effective way is to physically remove the front driveshaft and then test-drive the

If you look closely at this photo you can see several dimples in the tube of this driveshaft caused by trail damage. Even though the damage looks minor, the driveshaft's vibration was severe after it was hit on the trail. The only fix for this is having the driveshaft professionally re-tubed.

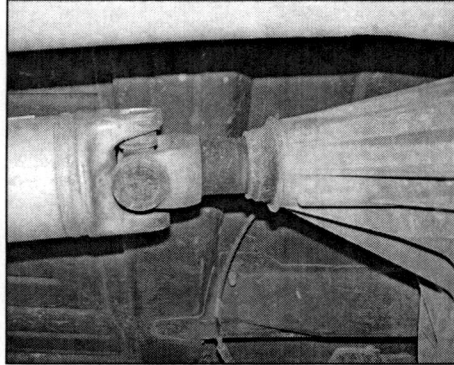

Many trucks have the slip joint for the driveshaft built right into the back of the transfer case as shown here. It is not uncommon for this type or even a normal slip joint to have some slack on higher-mileage vehicles. All it takes to check is grabbing the shaft and pulling up and down on it. If quite a bit of slack is felt, it is time for a replacement.

U-joints are another common wear item that can cause significant vibration if worn excessively. It is also common for a marginal U-joint that did not vibrate in stock form to suddenly start after a lift is installed. Also, since they are forced to operate at greater angles than factory, U-joints tend to wear more quickly on a lifted 4x4, especially one that is used frequently in muddy or dirty environments. As a result, driveshaft inspections should be done more frequently on a lifted truck.

vehicle to see if the vibration persists. If it does, then the problem is in the rear; if not, then the front shaft is the culprit. However, keep in mind that not all vehicles can be driven with the front driveshaft removed, so consult with your local off-road shop before driving without the front shaft.

Another good thing to check is that the transfer case is not stuck in full-time or four-wheel-drive mode. Remotely actuated transfer cases (i.e., push-button systems rather than a lever on the floor) are known to be problematic and don't always disengage when you tell them to do so. The procedure for verifying this varies, but if the front output on a part-time transfer case does not spin in two-wheel drive with the driveshaft removed, then the transfer case is stuck in four-wheel drive.

Physical Inspection

The easier of the two to diagnose, constant vibration can be traced to either worn or damaged

driveline components, or improper driveline angles. The first step is to crawl underneath and physically inspect the driveshaft for trail damage. Driveshafts hang low and are very susceptible to dings and dents from trail use. Also check to see if the slip-splines (in a driveshaft so equipped) have been twisted due to excessive torque being applied. Verify that the joints on the end of the shaft are perfectly parallel to one another. If they are off, even just a couple of degrees, serious vibration will result. If the driveshaft has been removed recently, verify the slip joint has been clocked properly to keep the joints on the ends of the shaft in phase with one another. If damage is found, fixing it will most likely eliminate the vibration.

While inspecting the driveshaft for physical damage, grab the shaft and pull it up-and-down as well as side-to-side to see if there is any play in the joints. This can be difficult to

detect without removing the driveshaft, but any movement at all indicates excessive wear. If the driveshaft has a CV, closely check to be sure the center knuckle is tight and free of any play. If a carrier or center support bearing is present, be sure it is serviceable. If some play is found, most likely that is the problem. Finally, it has been mentioned elsewhere, but bears repeating: a lift sometimes causes a marginal driveshaft to vibrate even if it did not do so in stock form. Verifying the 'shaft components are in good shape should always be the first step.

Angle Check

Once the front or rear shaft is identified as the problem and the shaft in question has been given a clean bill of health, checking for

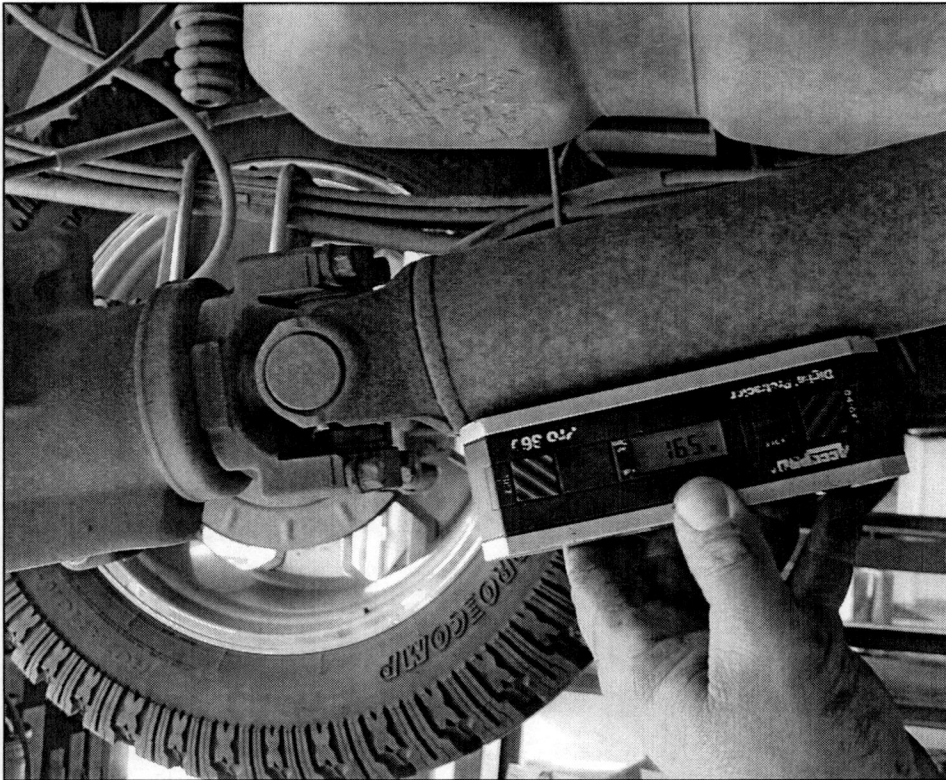

Though driveline correction is normally built right into a lift system's components, should there be a driveline problem it's a good idea to check the driveshaft's operating angles. Measuring the running angle of the driveshaft and the operating angle of the axle pinion and transfer case yoke quickly reveals any shortcomings. The angle finder shown is digital, but inexpensive analog versions are readily available.

If the driveshaft is equipped with a constant-velocity joint (CV) at the transfer-case end, the general rule is that the CV should operate at a maximum of 28 degrees (so much less than that should be present at normal ride height) and the pinion angle should be 1 to 2 degrees down from parallel with the driveshaft (once again at normal ride height). This 2-degree variance compensates for slight pinion lift under acceleration. As for the CV, be sure it is not binding internally ("knuckling out" in common parlance) at any point throughout the suspension's travel cycle.

If the driveshaft is equipped with conventional U-joints, the rule is that they should be running at equal but opposite angles to one another. This means if the transfer-case angle reads 8 degrees, the axle joint should measure the same in the opposite direction. Maximum safe operating angles vary among the experts and even among driveline manufacturers. Many manufacturers rate their U-joints at a very conservative 7 degrees, while 4x4 driveline experts rate conventional U-joints as high as 15 degrees. Keep in mind these numbers are maximums and there should be less present at normal ride height.

proper driveline angles is the next step. The specifications for this depends on the type of driveshaft in question and can vary a little among both driveline shops and lift manufacturers. Always follow the specific recommendations of the pros, but there are a few general guidelines that can be followed to check things out on your own. To do this you need an angle-finder, which is available in both analog and digital versions. The analog angle-finders are inexpensive and can be found at better tool stores.

Driveline angles increase as ride-height is raised, but they also increase more quickly on shorter-wheelbase vehicles than longer ones. This is why Jeeps are more prone to driveline problems than full-size trucks. CV driveshafts can handle greater angles for longer periods of time because their operating angle is divided between two U-joints, making them the common choice for lifted 4x4s. There are numerous resources available online that discuss the finer points of driveline angles as well as many driveline

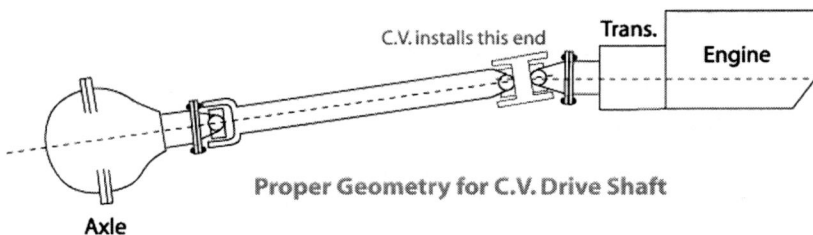

CV driveshaft operating angles.

shops that specialize in the unique needs of off-roaders.

Constant vs. Momentary

Identifying the type of driveline vibration the truck exhibits can often help identify the best fix. Constant vibration is most often caused by worn driveline components or angles that are way outside specifications. Vibration felt only under aggressive acceleration is usually the result of axle wrap. Vibration that comes and goes at a certain speed could be angularity or excessive wear. Constant vibrations should always be addressed, as should extreme momentary vibrations. Light vibration under certain conditions or certain speeds can be difficult to tune out and is sometimes a by-product of greater operating angles that fall inside maximum specifications. The fix for these can be nearly impossible and very expensive to find. Simply putting up with them or avoiding overly aggressive driving styles may end up being the best defense.

Driveline Fixes

Without exception, the goal for any lift manufacturer is to utilize the factory driveline components whenever possible and clearly identify when driveline alterations are necessary. As always, more lift makes drive-line issues more likely and there are exceptions to every rule. That said, the majority of the lift systems on the market work with existing driveshafts, and driveline correction is usually built right into the lift components. This makes driveline correction a no-brainer for most people, who often do not even notice that the angles have been "tweaked" to work with factory stuff.

Most taller lift blocks are not square but instead have a taper to them as seen with the block on the bottom. When positioned correctly with the thick end to the rear of the vehicle, the rear axle is tilted upward to relieve driveline angle. Shorter blocks are usually flat. Always check to be sure the blocks are positioned properly because installing them "backward" creates major driveline issues.

Rear driveline correction can take on a couple of different forms. In the case of lift blocks, the block face is often (but not always) tapered to rotate the rear-axle pinion upward. For this reason, it is important to be sure the blocks are installed properly, with the thick end of the block to the rear of the vehicle. On replacement springs, often the centering pin location (which positions the axle) has been moved ever so slightly to avoid having driveshafts lengthened. Degree shims are also common and serve the same purpose as tapered lift blocks: they move the axle pinion upward to relieve driveline angle. Always be sure the degree shim has been installed correctly (thick end to the rear on spring-over-axle applications and thick end to the front on spring-under-axle applications). These degree shims

Degree shims are the easiest way to correct driveline vibration with a replacement rear spring. Most spring manufacturers include shims in the design of a rear spring and ship them already installed so there's no guesswork for the installer. Shims are commonly available in two, three, and six degrees of correction.

come in a variety of thicknesses, so swapping them out can allow the pinion to be set to the proper angle if a problem develops. On driveshafts equipped with carrier bearings, brackets are usually supplied to space the bearing downward. Some of these spacers are even "tunable" to accommodate different wheelbase configurations. In a limited number of applications, a machined spacer can be installed between the U-joint and pinion flanges to avoid having a driveshaft lengthened.

In some situations, no amount of tweaking will allow the factory driveshaft(s) to survive, so a replacement is in order. These replacements usually allow a shaft equipped with a CV joint to be used in the place of a conventional U-joint-equipped shaft. In the case of most front and some rear driveshafts this is normally a remove-and-replace proposition, with the possibility of swapping out a transfer-case output yoke (a deceptively easy thing to accomplish). Note that replacement driveshafts are usually sold separately and can add substan-

Longer wheelbase vehicles typically have a rear driveshaft supported by a carrier bearing, and as you might guess it too needs to be adjusted when a lift is installed. In this case the bearing has been lowered with a selection of shims, so the installer can add or remove a few to "tune out" any vibrations. Carrier bearing spacer brackets are inexpensive but not always included with a standard lift system because not all trucks have them, so be sure your kit includes the appropriate bracket if your truck has a carrier bearing.

On the extreme end of the spectrum, replacement driveshafts are sometimes required for taller lift systems and some shorter ones on vehicles that use undesirable driveline components. Replacement driveshafts are typically longer than stock to accommodate the increased distance between the axles and transfer case, and sometimes they replace a conventional driveshaft with a CV-style shaft. This particular replacement shaft is for the front of an IFS Chevy. Always ask if new 'shafts are required when shopping for a lift system.

tial cost to the total expense of the lift system.

Jeep Wranglers manufactured between 1989 and 2006 have a special option available. The non-Rubicon models are equipped with an NP231 transfer case, which is plenty strong and capable of surviving behind V-8 power. Their only weakness is a slip-style rear driveshaft. For several years now Slip-Yoke Eliminator (SYE) kits have been available to fix several problems at once: They lengthen the driveshaft, allow a CV style driveshaft to be used, and allow the vehicle to still be driven in the event of a drive-line failure. SYE kits are advantageous for 4-inch lift systems and required for most 6-inch systems. Though they require a bit of an investment, they are really mandatory if the Jeep is used off road more than occasionally. Many Cherokees from '89–2001 can also benefit from an SYE kit.

Stock Jeep NP231s come equipped with an excessively long and failure-prone rear output shaft. Designed to work with a slip-yoke driveshaft, this is not a great design for off-road use. Fortunately, these transfer cases can be beefed up significantly to handle both more horsepower and lift height.

No More Slip

Excessive rear driveline angles have been a problem for Jeeps since people first started lifting them. In stock form, short-wheelbase Jeeps have a driveshaft length of only 13.5 inches or less. Because of the short length, driveline angles increase very quickly as the Jeep is lifted. Conventional correction methods can only do so much, and once a lift height of around 6 inches is reached, the only real fix is a longer driveshaft. Jeep CJs had it bad, but once the NP231 transfer case appeared in 1989 under the YJ, something had to be done.

Enter the Slip-Yoke Eliminator (SYE) kit. This relatively simple modification replaces the factory slip-yoke driveshaft with a sturdier and more reliable CV driveshaft. This conversion does several things at once: it increases driveshaft length about 4.5 inches (6.5 inches on TJs), enables the use of a CV joint that is capable of operating at greater angles, and prevents transfer case fluid loss in the event of a rear driveshaft failure (a huge deal for off-road enthusiasts). The extra length flattens out the angle of the driveshaft, which can mean the difference between lowering the transfer case (which hampers ground clearance) and even allowing the driveshaft to survive at all.

These kits are available in a variety of different forms, but one of the better SYE kits is available from JB Conversions. This kit includes a high-quality mainshaft, a new aluminum tailhousing, a few snap rings, and a new yoke. Everything needed is supplied with the exception of a new driveshaft, which varies in length depending on the application. Any reputable driveline shop can handle building the new, stronger, and longer driveshaft. For extreme applications, JB Conversions also offers a super-short SYE kit with another 3⅛ inches of driveshaft length over its standard kit, making the overall length of the 231 shorter than the venerable Atlas transfer case.

Installation requires partially disassembling the transfer case, but as luck would have it, NP231s are dirt simple and nothing special beyond a pair of sturdy snap ring pliers is required. The transfer case doesn't even have to be removed from the vehicle, though the process is a little easier on a bench. Still, diving into the guts of a transfer case can be scary to many people, so to show you just how easy it is, the author performed this surgery himself on a YJ transfer case (TJ and Cherokee 231s are virtually identical in both parts and installation steps). The whole process took under two hours including photo opportunities. In the end the Jeep had benefited from a stronger 'case and all the length needed for the driveshaft to live with a 7-inch lift.

1. After removing the transfer case from the Jeep and draining the fluid, remove the bolts for the oil seal and gently tap it off of the tailhousing. Behind it is the first of several snap rings that need to be removed. This one is wrapped around the mainshaft and holds the rear output shaft bearing in place. (Tom Morr)

2. Next remove the speedometer drive gear making careful note of how it is clocked in the tailhousing (this is important for re-assembly). Stand the transfer case on end and support it with a block of wood. Remove the bolts securing the tailhousing to the back half of the case and gently pry it upward. This can be tough because it is "glued" in place with silicone, but use caution not to damage the back half of the case.

No More Slip *continued*

3. Next remove the bolts securing the back case half and gently pry it apart (once again, you'll be fighting a sticky silicone seal). Once the seal is broken, carefully lift the case half off of the main body, taking precautions not to damage the oil pump that is now visible on the back of the case.

4. At this point you're staring straight at the guts of the case, but don't worry because there's really not much to it. Remove the nut on the front output yoke and tap it off of its shaft. This yoke and nut will be re-used.

5. Slide the spring off of the end of the shift fork and put it someplace where it won't get lost. Now grab the front output and the mainshaft and pull it upward with the chain and shift fork. This should all come out as an assembly and it should slide out of the case fairly easily.

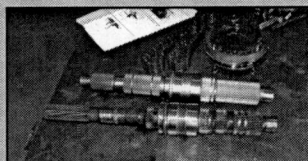

6. Remove the chain from the mainshaft and set it aside along with the front output. Next, remove the large snap ring holding the sprocket hub to the mainshaft. Once again, this snap ring is strong and it may take a few minutes to get it out of its groove. Once out, there may be some needle bearings inside the bore of the hub. If present, they need to be pressed out because later 231s do not use them.

7. When comparing the old and new mainshafts, the difference in length becomes obvious. Also note the original mainshaft (the lower one in this photo) has some spiral cut splines for the speedometer drive gear. This is a well-known weak point in the original design and it is not uncommon to see the factory shaft fail in the spiral-cut area. The new shaft does not have the spiral gears and also accommodates a 32-spline yoke, making it considerably stronger than the factory shaft.

8. Slide the sprocket hub on to the new mainshaft making careful note that the hub ring (where the shift fork rides) is installed as shown, with the "thick" end toward the end of the mainshaft that will accept the rear output yoke. It can be installed backwards, which causes the unit to not function properly. Install the supplied snap ring to retain the hub and be sure it's fully seated in its groove.

9. Before installing the new mainshaft, clean the gasket surfaces on the case halves. Avoid the urge to use a sharp knife or a wire brush as this damages the soft aluminum gasket surface. It's also a good idea to clean the magnet (near the installer's left hand in a pocket built into the side of the case) of any debris. Of course, lots of metal debris on the magnet means it is time for a complete rebuild.

No More Slip *continued*

10. Aside from the snap rings, the only tricky part to this installation is sliding the new mainshaft in place. Just like during removal, the mainshaft, chain, shift fork, and front output shaft should all be installed as one assembly. It can be helpful to have an extra set of hands for this step to help guide the shift fork into its hole in the front case.

11. Apply some silicone to the gasket surfaces on the case halves. Verify the pickup tube is properly seated in the oil pump on the back case half then gently slide the pump and case half into place. Reinstall the bolts and torque them in a star pattern.

12. Now slide the plastic speedometer drive gear onto the main shaft and gently seat it on top of the oil pump. Note a "thick" and "thin" smooth section on either side of the spiral gears; install the gear with the thin end up as shown and secure it with the supplied snap ring.

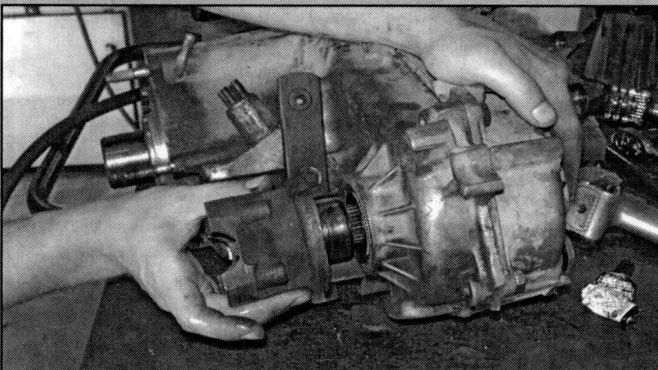

13. Apply a bead of silicone to the case and install the new tailhousing supplied on the back of the transfer case. The rear output bearing is already installed along with the oil seal. The tailhousing only installs one way, so it's a no-brainer. Use the factory bolts and tighten the housing in a star pattern.

14. Lubricate and install the original front output yoke on the output shaft, then secure it using the supplied star washer and original nut.

15. After installing the speedometer drive gear (make sure it clocked the same way it was before), this transfer case is ready for installation. Once it's in, lower the Jeep to the ground and simply measure for a new driveshaft. Thanks to the SYE kit, the transfer case is capable of holding up behind V-8 power, and the longer driveshaft makes all the difference with the 7-inch lift going on the recipient of this modified 'case.

CUSTOM AND HIGH-PERFORMANCE SUSPENSION SYSTEMS

Being able to handle big air requires a highly capable suspension system. Anyone can get big air, but being able to stick the landing is a whole different story.

Up until now the main focus of this book has been production aftermarket suspension systems that are designed to enhance factory designs. All of those systems work well and are ideally suited for a daily driver, a combination work/play truck, and for vehicles that serve primarily as week-

end toys. But what if bolt-on stuff is not enough? What if the ability to drive down the road is secondary to your dream of tackling nearly impossible terrain and generally pushing the envelope? If that's the case, it's time to break out the torch and welder, because this generally means

delving into a whole new world of high-performance suspension design where extensive modifications are the rule rather than the exception.

This chapter is a primer on the different types of high-performance suspension technology out there. Some of this stuff is available from select suspension companies, while other items come directly from racing and other competition fab shops. Building your own high-performance suspension system is a subject that can fill a book, but perhaps this provides a glimpse at what possibilities exist for your own off-road vehicle.

Ready-Made High Performance

Bolt-on suspension companies focus primarily on their bread and butter, which is easy-to-install lift systems that provide more ground clearance and room for larger tires. However, a select few also develop products that take suspension design to the next level. These are largely purpose-built designs that greatly enhance the capabilities of a vehicle, but often this means extensive modifications that

Moderate-to-hard trail duties can be handled with most off-the-shelf systems, but being able to tackle the hard stuff gracefully (like the Moab Rim in Utah where this photo was taken) and take on really radical terrain often means stepping up to more extensive suspension modifications.

take quite a bit of time and money to install. Usually these systems require extensive modifications to other vehicle components, including the steering, exhaust, and transfer case. These "extras" can sometimes double the overall cost of the project. Still, many off-road enthusiasts do whatever it takes to build a vehicle capable of tackling the toughest terrain.

Long-Arm Systems

Much has already been said about the advantages of a long-arm kit for TJs, and thanks to their enormous popularity, a few of the large suspension companies and many of the smaller ones now offer long-arm conversions. These involve replacing the short factory arms with much longer ones that extend all the way to the center of the Jeep, attaching to a new transmission crossmember. The operating angles of the longer arms change less as the suspension cycles when compared to the factory arms, so more suspension movement is possible before

the arms begin to bind within their mounts. Long-arms also enable less caster change. This is because the front axle moves and can accommodate more ride height than traditional short-arm systems. The bottom line to all of this is more suspension travel and a more controlled way to run tires above 35 inches tall.

While long-arm systems are great, they do have their drawbacks. They are very complex and require quite a bit of bracketry, so they are much more expensive than a traditional lift (usually more than double). They also require removing some or all of the factory suspension mounts from the frame, which is a time-consuming process that makes it very difficult (if not impossible) to reinstall the factory suspension later on. They usually require new driveshafts to handle the additional suspension travel, and some are more compatible than others when it comes to working with other vehicle upgrades like bigger axles and engine swaps. Still, once installed, a long-arm-equipped TJ is hard to beat for a serious trail machine.

Coil and Coil-over Conversions

Leaf springs are simple and can be made to perform well in just about any environment. Even so, there's

Long-arm systems provide substantial performance improvements, but are complicated to install. Note the factory suspension attachment points have been removed from the frame. The lower arms on this Superlift system have also been bent to maximize ground clearance.

The original Black Diamond XCL system was revolutionary when it was introduced in the mid '90s and the design holds up well today. Coil-overs at all four corners makes for an impressive amount of flex, but all that extra suspension travel comes at the expense of some stability. Here the author is getting some first-hand experience with the XCL in Moab.

The X2 continues to utilize a coil-over shock up front (though with updated valving and springs) and modernizes the look, but using a tubular shock hoop rather than the XCL's plate steel tower. The suspension link design and geometry has also been completely revised to improve performance and durability.

only so much you can do with a leaf spring before putting a coil spring in its place becomes an attractive alternative. Shade tree mechanics have been doing coil conversions on their leaf-sprung Jeeps for decades, but in recent years a few companies have offered conversion kits that often combine long-arm technology.

Black Diamond was the first company to offer a coil-over conversion for Jeep YJs in the mid '90s. The Xtreme Coil Link (XCL) system was an instant hit among the hardcore crowd with good reason: the XCL ditches leaf springs and instead utilizes a triangulated four-link for the rear and a radius-arm style front with coil-over shocks at all four corners. Though the modifications needed to the frame are extensive, the XCL is still popular for pure hardcore rock crawling and trail use. Black Diamond has since been sold to Superlift, who still offers the XCL for sale, but has recently released an

updated version called the X2. The X2 combines the experience and technology that Superlift utilizes with their long-arm TJ systems while addressing many of the shortcomings of the original XCL. The X2 utilizes a spring-over conversion for the rear that greatly enhances overall sta-

bility without sacrificing suspension travel thanks to a uniquely designed spring. It also utilizes a traction bar called a Torque Fork to control axle

The rear of the X2 maintains the use of leaf springs primarily to enhance stability, though it does include new springs designed for use with a spring-over. An optional traction bar is also available and is recommended for vehicles with V-8s and low gearing.

Skyjacker's Z-Link conversion for leaf-spring Ford Super Dutys is the first of its kind. It utilizes the factory spring hangers with new link arms and a coil-over shock. Skyjacker also offers its Platinum Series coil-over conversions for late-model Dodge Rams, Jeep TJs, and coil-spring Super Dutys. (Courtesy Skyjacker)

wrap while still allowing the rear suspension to flex. Up front, a long-arm-style four-link connects to the frame via a unique belly pan. This four-link connects to the front via bolt-on or weld-on brackets that essentially convert the front axle to a TJ. The actual suspending duties are handled by coil-over shocks with 10 inches of travel. Predictably, the X2 takes extensive modifications to the frame and axles, but Superlift has also put five years of design work into the system in order to be sure all

of the bugs were worked out.

Other companies further up the ante on TJs with front coil-over conversions. Usually offered in conjunction with their long-arm systems, replacing the coils with a race-type coil-over shock enables better spring rates and more precise shock valving. Skyjacker's Platinum Series coil-overs are proven performers, while Superlift utilizes much of the X2 design on its coil-over conversion. Fabtech and Full Traction also offer a TJ coil-over option.

More recently there has been a trend toward applying coil-overs to larger trucks, such as the new Super Duty, Dodge Ram, and other models. When used in conjunction with a suspension lift, these coil-over conversions offer pre-runner capability right out of the box.

Tube Chassis Tech

There comes a point for many off-road-racing competitors and some rabid trail enthusiasts where it makes more sense to build a purpose-built chassis designed to individual specifications than it does to make radical modifications to a production vehicle. Enter the tube buggies or tube chassis. These scratch-built chassis throw sheet metal out the window or at the very least make it a secondary consideration. When starting from scratch, virtually anything can be built to suit the needs of the owner, whether it's a buggy intended for hardcore trails or a chassis for competing in one of the many forms of off-road racing.

In the '90s if you wanted a tube-chassis off-roader, your options were either to build it yourself or go to a race shop where one could be built to your specifications. Today there are several companies that offer tube-frame chassis from bare all the way up to turnkey machines. One of the companies at the forefront of the ready-made chassis business is Poison Spyder Customs. The company's Bruiser chassis is available in both two- and four-seater versions and with any number of options to suit an individual's needs and budget. The chassis is designed to look more or less like a production Jeep, and the end result is a distinctive style that performs amazingly well in just

Poison Spyder Custom's Bruiser chassis is one of the most popular spec-built off-road chassis for hardcore trail use, and PSC has sold a bunch of them. With all of the difficult design work done for you, a chassis such as this is hard to beat when building a hardcore trail machine. (Courtesy PSC)

The majority of PSC's Bruiser chassis are shipped as you see here; a roller with the suspension work and axles completed. This way the owner can add whatever drivetrain is preferred and add other finishing touches to the chassis. (Courtesy PSC)

In completed form a properly built Bruiser is a force to be reckoned with on any hardcore trail. This one is called Suicide Sally and it serves as a calling card for PSC at shows and on trails. This level of craftsmanship extends to the rest of PSC's product line. (Courtesy PSC)

and capable platform for a hardcore off-roader.

Other Suspension Designs

Though shade tree mechanics constantly tinker with all kinds of different suspension designs, most scratch-built systems end up using a four-link system to locate the axle with a coil-over or air shock serving as the weight-supporting member. Four-links come in a variety of different styles. There is the conventional four-link that uses a track bar as seen on many production vehicles (though technically this is a five-link). There are also triangulated four-links that eliminate the need for a track bar. With a triangulated system, the upper, lower, or both sets of suspension links are mounted at an angle, thus locating the axle up-and-down as well as side-to-side. Fabricating a four-link system is very difficult for a novice and

A variation of the full tube chassis, PMP Performance offers a TJ buggy kit in which a tube chassis is built to attach to a factory Jeep TJ frame. The unique thing about this kit is that many of the donor vehicle's components can be retained (which saves substantial money) and the tube chassis provides substantial weight savings. Any bolt-on TJ suspension system can be used or the owner can hand-fabricate one as well.

about any terrain. As a leader in this small but rapidly growing market, the company has sold many copies of its chassis over the years.

So why a section about tube chassis in a suspension book? Because with a tube chassis, you can design any kind of suspension you want. There are no frame rails or floorboards to get in the

way, and most tube-frame builders admit that suspension design and geometry are a primary consideration when building a frame to spec. Another major advantage to a tube chassis is weight savings; believe it or not, tube-frame vehicles usually end up weighing far less than a production vehicle, and less weight means a more stable

---

Four-link systems have become the standard among competition and hardcore trail rigs today. Though they look simple, there is quite a bit of engineering behind one that performs as it is supposed to. Note how both the upper and lower links are angled between the chassis and the axle; this triangulation serves to locate the axle side-to-side rendering a track bar unnecessary.

involves quite a bit of math as well as engineering know-how in order for a four-link to work properly. In other words, there are a whole bunch of ways to mess up a four-link and only a couple ways to get it right. For this reason, four-link fabrication is best left to the professionals.

One rather unique alternative to coil springs, leaf springs, and coil-over shocks is a quarter-elliptic system. Imagine a leaf spring cut in half and then turned upside down with the eye attached to the axle and the body of the leaf pack at the frame. Quarter-elliptics are a cheaper alternative to coil-over shocks and can be made to accommodate more flex than a conventional coil spring, but some sort of link system is still needed to locate the axle properly. Quarter-elliptic designs are not all that common any more, but when built properly they are capable of coil-over performance at a fraction of the price.

Fabricators experimented with quarter-elliptic springs extensively when rock crawling competitions were just starting, and a few can still be seen today. A viable alternative to costly coil-over shocks, quarter-elliptics still require quite a bit of know-how to set up correctly. (Courtesy PSC)

Racing Inspiration

When contemplating different custom-suspension designs, look no further than the many different forms of racing for ideas. Four-link theory was not invented by off-roaders and neither were tube chassis or coil-over shocks; these things and many others were all adapted from many different forms of motorsports. Getting ahead in the racing world usually requires thinking outside the box, and many of the cutting-edge technologies used in today's suspension designs, both factory and aftermarket, were bred in some form of racing. What will the next big trend to come from racing be?

The Domino Effect

It was mentioned in Chapter 1 but especially applies to high-performance suspension systems: beware the domino effect. More often than not, high-performance systems advance the vehicle's capability well beyond other factory parts. Being able to run 37-inch tires with 14 inches of suspension travel is good, but to maximize these gains, lower gears and locking differentials are needed. But then those items most likely overtax the factory axles; so custom-built axles are needed, which are heavier than the stock pieces. Now with all the extra weight, the

Baja and short-course off-road racing are the pinnacles of technology in the off-road world today. A multitude of long-travel two- and four-wheel-drive suspension designs are used in both types of racing that provide upwards of 20 inches from full extension to full stuff. Taking a close look at these race trucks can stimulate all sorts of ideas that can be applied to everyday trail rigs.

stock engine just doesn't want to perform like you want, so an engine transplant is in order, which puts out too much horsepower for the stock transmission to handle, so... you get the idea. But hey, isn't tailoring a vehicle to do what you want part of what off-roading is all about?

SUSPENSION GLOSSARY

With a topic as complex and broad as "off-road suspension," it might be helpful to have a clear definition of the many technical terms used throughout this book. Special thanks to Superlift, who provided excerpts from the glossary used in the company's catalog, upon which some of these definitions are based.

Ackerman angle: This angle defines the correct angle that steering wheels must have to effeciently negotiate a turn. In a standard turn, the wheel on the inside of the turning radius must follow a more severely curved line than the outside wheel. By setting steering arms, one can successfully adjust the turning angle so that the inside tire turns at a larger angle than the outside. This results in better traction around the turn.

Anti-sway bar: See "sway bar."

Articulation: How well each axle, or each wheel, travels and "twists-up" when traversing an off-camber obstacle. When off-roading, you want as much suspension articulation as possible to keep the tires in contact with the ground (traction). Suspension manufacturers typically strive to achieve the right balance of off-road articulation and on-road stability.

Axle Wrap: A condition in which as power is applied, torque load causes the rear pinion to pivot upward. This load causes the leaf springs to deform from their natural shape and apply force in the opposite direction. At some point these forces overcome tire traction, which induces wheel-hop. The "fix" is to increase spring strength, reduce leverage by decreasing lift block height, and/or install traction aid devices. Axle wrap can

happen on either front or rear solid axles.

Bumpsteer: This can be caused by a number of alignment related issues. This is when a vehicle darts or wanders while steering input remains unchanged. Usually encountered when the vehicle is operated on a less-than-ideal driving surface. In other words, it takes a concentrated effort to keep the vehicle in a straight line.

Bump stop: See "Compression travel."

Body roll: Experienced when cornering or during emergency maneuvers, body roll is most noticeable when turning at speed. One side of the suspension compresses as the opposite side lifts, causing the passenger compartment to lean towards the outside of a turn. Sway bars are designed to limit the amount of body roll in a suspension system. Though more noticeable on heavy vehicles such as SUVs, body roll can happen on any vehicle.

Camber angle: An alignment term related to steering geometry. Camber is the inward or outward tilt of a tire in relation to the bottom when compared to a vertical line. For a complete explanation, refer to Chapter 2. Camber is impacted the greatest when lifting a Ford with TTB suspension, but IFS systems can also suffer from camber problems with a lift system.

Caster angle: An alignment term related to steering geometry. Caster is the "forward" or "rearward" tilt of the front suspension balljoints as compared to a vertical line. Caster has a great deal of impact on the steering's ability to self-center and a variety of other driveability traits. See the full discussion on caster in Chapter 2.

Centerlink: The portion of the steering linkage that connects the pitman and idler arms, commonly found on IFS vehicles that have upper and lower control arms. A tie rod connects to each end of the centerlink. Depending on lift method, the factory centerlink is retained, relocated, or replaced entirely.

CNC: Acronym for Computerized Numerical Control. Term describes a type of control system used on a piece of manufacturing equipment. CNC machines offer unsurpassed accuracy and repeatability.

Coil-over shocks: A broad term applied to similar components that may look very different. A coil-over is made up of a coil spring and shock that are designed as one integral unit and which also use the same mounting points. Also called "struts" and used on many late-model applications, factory struts are not quite on the same level of looks or performance as the coil-over shocks used in various forms of racing. The basic construction is a coil spring wrapped around a shock, with the main variation being that factory struts are not adjustable, while race-inspired coil-overs are. Both versions are used extensively in the aftermarket.

Compression travel (stop): A measurement of the amount the suspension compresses before it bottoms-out against its travel-stops. This travel-stop is also called a "bump stop" and "jounce stop."

Control arms: Also called "A-arms" because of their shape, control arms are found on all IFS suspensions (discussed in Chapters 6 and 7). There is one upper and one lower arm on each side. Control arms have a balljoint on the outboard end, and connect to the frame on the inboard end.

CSS: Acronym for Centerlink Stabilizing System. A dropped centerlink tends to pivot fore and aft excessively when turning force is applied. The CSS uses one or two links to tie the centerlink to a crossmember and prevent this excessive movement.

Curb weight: How much the vehicle weighs when loaded with a normal compliment of passengers, fuel, gear, etc.

CV axle: Acronym for Constant Velocity axle. With control-arm-style IFS, they are the rubber-booted axle shaft assemblies (one per side) that bolt to the differential housing flange on the inboard end, and mate to the hub/knuckle on the outboard end.

DOM: Acronym for Drawn Over Mandrel. The term specifies a certain type of tubular steel that has exceptional strength, forming, and welding characteristics.

DOT: Acronym for (Federal) Department of Transportation.

Drag link: A piece of steering linkage that connects to a pitman arm on the upper end and a tie rod or knuckle on the lower end. On lifted vehicles, a dropped pitman arm often reduces the Original Equipment (OE) drag link operating angle. In the case of leaf-spring Chevy, Dodge, and Toyota trucks, the factory drag link is replaced by a "dropped" drag link to reduce link angle.

Driveability: The sum of the vehicle's driving traits and mannerisms. Handling, steering traits, and ride quality are the major categories that influence driveability. Driveability is an important concern for most lift designers and it should be equally important to potential lift system buyers.

Driveshaft angle: The operating angle of the driveshaft in relation to the pinion yoke (at the differential) or the output yoke (at the transfer case). This angle is greatly impacted by driveline length on lifted vehicles, and greater angles

mean greater risk of developing driveline vibration or bind.

Drop Pitman Arm: A form of steering correction commonly used with lift systems, these arms have more distance between the sector shaft and drag link attachment points to relieve drag link angle and correct the steering to work with a lift system.

Extension travel: A measurement of the amount the suspension extends before it tops-out against its travel stops.

Four-link: A suspension type that uses two upper and two lower link-arms to connect the solid axle (front or rear) to the frame. The Jeep TJ has a common four-link system.

GVWR: Gross Vehicle Weight Rating. This information is provided by the vehicle manufacturer and is located on a tag generally found on the driver-side doorjamb. It tells you how much total weight, including occupants, fuel, bed load, etc., the vehicle is rated to carry.

Idler arm: Found on IFS vehicles that have a centerlink, the idler arm supports one end of the centerlink while the pitman arm supports the other.

IFS: Independent Front Suspension. With this type of suspension, the wheels travel independently of each other. The IFS vehicles covered in this book are either control arm types or Ford TTB.

Knuckle: A casting attached to the outboard ends of the axles that pivot on upper and lower ball joints. The knuckles serve as the attachment points for the spindle/hub/wheel assembly. Replacement knuckles are used on lifted IFS vehicles when the lower control arms are lowered but the upper control arms are not.

Lift System: A complete system or kit to add ride height to a vehicle. For some this term indicates a lift that includes replacement springs, but to most people "lift kit" and "lift system" are synonymous. The terms are used interchangeably throughout this book.

Military wrap: Featured in some leaf spring designs, the second leaf plate wraps around the spring eye (called the main leaf) to form a double-wrap. This design element transmits less

stress to the main leaf during extreme articulation. It also lessens the odds of spring separation in case of main leaf failure, and provides greater strength, support, and durability as a whole.

OEM (or OE): Acronym for Original Equipment Manufacturer. In this book the OE is the actual vehicle manufacturer such as Ford, GM, etc.

Panhard bar: See "Track bar."

Pitman arm: This steering component has splines that engage on the steering sector output shaft. The opposite end connects to a drag link or centerlink, depending on steering system design. With a dropped pitman arm, the drag link attachment point is moved down to reduce link angle.

Pre-set coil springs: Pre-setting involves fully compressing the coil spring (so that the coil wraps are actually touching each other) in the manufacturing process. This reduces the amount of coil sag and extends coil service life.

Progressive rate coils: Most coil springs have rates that are constant; if it takes 400 pounds to compress the spring the first 1 inch, another 400 pounds will compress it the second inch, and so on. With a progressive rate coil, the initial couple of inches of compression require less spring rate than the remainder of compression. This design allows good ride characteristics, but also has the increased rate necessary to handle rough terrain. With progressive rate coils the first few coil wraps are more closely spaced than the remaining coils, or the entire spring is slightly conical (cone-like) in shape.

Radius arm: The arms (one per side) run basically parallel with the frame rails and locate the front axlehousing to prevent fore and aft axle movement. Common vehicles factory equipped with radius arms are 1966–'79 Fords with coil spring/solid axle suspension, 1980–'96 Fords with coil spring TTB, and 2005 and newer Super Dutys. Caster angle must be addressed when lifting vehicles with radius arms.

Roll and yaw: The amount of sideways lean and front-to-rear oscillation felt on a vehicle, most noticeable on SUVs. The sensation compares to the side-to-side oscillations of a boat. The problem exists on lifted vehicles that have incorrect track bar length and/or severe track bar angle. When lifted, the track bar needs to be lengthened or binding occurs as the suspension travels; if unaddressed, the frame and axlehousing continuously "tug" against each other. This is most noticeable when the vehicle is being driven in a straight line, going through highway dips at speed. A certain amount of "steering wheel kick" normally accompanies roll and yaw.

Solid axle: A one-piece axlehousing design that has rigid axle tubes all the way out to the knuckles. Good examples of vehicles with solid front axles are 1966–'79 Fords and 1969–'87 GM trucks.

Spring rate: A measurement of force (in pounds) required to compress a spring a given distance (in inches). Be careful when comparing rates since all manufacturers do not use the same measuring procedures.

Steering arm: This forged steel component bolts to the front axle knuckle. Its opposite end attaches to the drag link. A raised steering arm is taller to reduce drag link angle.

Steering wheel kick: This problem is most noticeable when driving in a straight line and the suspension compresses, like when going through a highway dip at speed. The steering wheel moves slightly, but the vehicle continues to track straight ahead. If the vehicle has a track bar, the problem is related to incorrect track bar length (too short) and/or incorrect phasing. "Incorrect phasing" exists when the track bar arc-of-movement is not in phase with the drag link arc-of-movement. Incorrect phasing can result when a dropped pitman arm is used, but the track bar is not lowered, or vice versa. A certain amount of mismatched phasing is acceptable, but at some point "kick" occurs.

Strut: See "Coil-over shock," though usually strut is used in reference to factory suspension components.

Stud bind: When a tapered stud, normally a tie rod or track bar end, over-extends its pivot capability. This is a byproduct of excessive operating angles.

Suspension articulation: see "Articulation."

SUV: Sport Utility Vehicle. These are multi-passenger wagon-type vehicles usually based on truck platforms, not to be confused with the latest crop of crossover utility vehicles that are usually based on car platforms.

Sway bar: A roughly U-shaped tube that attaches to the frame rails and each side of the suspension. Under cornering conditions, a sway bar can take suspension compression on one side and introduce preload on the opposite side, thus reducing the roll and yaw of a vehicle. Sway bars are present on most front suspensions as well as selected rear suspension designs. Like everything else, sway bars must be addressed with a lift system.

Sway bar links: These connect the sway bar (attached to the body) to an anchor point on the axle or control arm. Generally, the bar body is relocated or longer links are used to compensate for a lift system.

Sway bar pre-load: As lift height increases the sway bar body pivots and begins to pre-load. This pre-load puts extra stress on the links and reduces suspension flexibility. Proper bar geometry is restored by relocating the body or by using longer links.

SYE: Slip-Yoke Eliminator. This is a kit that shortens the overall length of the transfer case, therefore increasing driveshaft length. An SYE kit allows the use of a CV driveshaft that is capable of handling greater operating angles than a conventional driveshaft. SYE kits can be installed on Jeep YJs, TJs, and Cherokees equipped with an NP231 transfer case.

Tie rod: A section of steering linkage. Though a tie rod can be configured a number of different ways, a tie rod always attaches to at least one steering knuckle. Tie rod ends are wear items that should be checked periodically, especially on lifted vehicles.

Tie rod boss: The portion of the steering knuckle where the tie rod attaches. IFS knuckle systems typically use a knuckle that has a raised tie rod boss when compared to the factory knuckle. This is done to correct the operating angles of the steering, but some knuckle designs may quicken the steering ratio and/or reduce turning radius depending on the tie rod boss location.

Toe angle: A steering geometry term; the side-to-side difference in distance between the front and rear of the front tires. If the distance is closer at the front, it's called toe-in. If the difference is closer at the rear, it's called toe-out. See Chapter 2 for the full discussion.

Track bar: A bar that runs laterally (perpendicular to the frame) from the frame to the axlehousing. Found on almost all coil-spring-equipped rigs that have solid axles, and some leaf-spring-equipped vehicles. The bar helps locate the axlehousing side-to-side and is a critical factory in overall suspension stability with coil spring designs. Generally, when lift exceeds 2 inches, the bar must be relocated and/or replaced with a longer, adjustable unit.

Track width: The measurement from outside of tire to outside of tire. Track width increases when wider tires and/or wheels with less backspacing are installed. Most IFS knuckle systems increase track width.

TTB: Twin-Traction Beam, Ford's Independent Front Suspension system found on most 1980 to 1996 models. See Chapter 5.

Turning radius: A measurement of the distance required to turn a vehicle.

Wheel hop: The violent jumping up-and-down of the tires (usually on the rear axle) caused by axle wrap. The "jumping" is actually the tires catching and then releasing traction with the ground, often several times a second.

Wheel travel: See "Suspension travel."

Wrap-up: See "Axle wrap."

Source Guide

Bilstein
14102 Stowe Drive
Poway, CA 92064
Phone: 858-386-5900
www.bilstein.com

Black Diamond
260 Huey Lenard Loop Rd.
West Monroe, LA 71292
Phone: 866-680-6666
Fax: 318-397-3040
www.blackdiamondoffroad.com

Edelbrock Corp.
2700 California St.
Torrance, CA 90503
Phone: 310-781-2222
Fax: 310-320-1187
Tech Line Only: 800-416-8628
www.edelbrock.com

Explorer Pro Comp
Phone: 800-776-0767
www.explorerprocomp.com

Fabtech Motorsports
4331 Eucalyptus Ave.
Chino, CA 91710
Phone: 909-597-7800
Fax: 909-597-7185
www.fabtechmotorsports.com

Fox Racing Shox
130 Hangar Way
Watsonville, CA 95076
Phone: 800-FOX-SHOX
Fax: 831-768-9342
www.foxracingshox.com

Full-Traction Suspension
6951 McDivitt Dr.
Bakersfield, CA 93313
Phone: 800-255-6464
Fax: 661-398-9555
www.fulltraction.com

King Shock Technology, Inc.
12842 Joy St.
Garden Grove, CA 92840
Phone: 714-530-8701
Fax: 714-530-8702
www.kingshocks.com

Nth Degree Mobility
44 Miles Rd.
Mound House, NV 89706
Phone: 775-885-8454
Fax: 775-885-8422
www.nthdegreemobility.com

Rancho
1 International Dr.
Monroe, MI 48161
Phone: 734-384-7804
www.gorancho.com

Rock Krawler Suspension
Phone: 518-270-9822
www.rockkrawler.com

Rough Country Suspension Systems
1400 Morgan Rd.
Dyersburg, TN 38024
Phone: 800-222-7023
www.roughcountry.com

Rubicon Express
3290 Monier Circle #100
Rancho Cordova, CA 95742
Phone: 877-367-7824
Fax: 916-858-1963
www.rubiconexpress.com

Skyjacker Suspensions
212 Stevenson St.
West Monroe, LA 71292
Phone: 318-388-0816
Fax: 318-388-2608
www.skyjacker.com

Superlift Suspension Systems
300 Huey Lenard Loop Rd.
West Monroe, LA 71292
Phone: 800-551-4955
Fax: 318-397-3040
www.superlift.com

Sway-a-Way
20724 Lassen St.
Chatsworth, CA 91311
Phone: 818-700-9712
Fax: 818-700-0947
www.swayaway.com

Tera Manufacturing, Inc.
5251 South Commerce Dr.
Murray, Utah 84107-4711
Phone: 801-288-2585
Fax: 801-713-2313
www.teraflex.biz

Trail Master Suspension
a division of Performance Automotive
Group, Inc.
3651 N. Highway 89
Chino Valley, AZ 86323
Phone: 928-636-3175
Fax: 928-636-7079
www.trailmastersuspension.com

Tuff Country EZ-Ride Suspension
4172 West 8370 South
West Jordan, UT 84088
Phone: 800-288-2190
Fax: 801-280-2896
www.tuffcountry.com

MORE GREAT TITLES AVAILABLE FROM CARTECH®

CHEVROLET

How To Rebuild the Small-Block Chevrolet* (SA26)
Chevrolet Small-Block Parts Interchange Manual (SA55)
How To Build Max Perf Chevy Small-Blocks on a Budget (SA57)
How To Build High-Perf Chevy LS1/LS6 Engines (SA86)
How To Build Big-Inch Chevy Small-Blocks (SA87)
How to Build High-Performance Chevy Small-Block Cams/Valvetrains (SA105P)
Rebuilding the Small-Block Chevy: Step-by-Step Videobook (SA116)
High-Performance Chevy Small-Block Cylinder Heads (SA125P)
How To Rebuild the Big-Block Chevrolet* (SA142P)
How to Build Max-Performance Chevy Big Block on a Budget (SA198)
How to Restore Your Camaro 1967–1969 (SA178)
How to Build Killer Big-Block Chevy Engines (SA190)
Small-Block Chevy Performance: 1955-1996 (SA110P)
How to Build Small-Block Chevy Circle-Track Racing Engines (SA121P)
High-Performance C5 Corvette Builder's Guide (SA127P)
Chevrolet Big Block Parts Interchange Manual (SA31P)
Chevy TPI Fuel Injection Swapper's Guide (SA53P)
How to Rebuild & Modify Chevy 348/409 Engines (SA210)

FORD

High-Performance Ford Engine Parts Interchange (SA56)
How To Build Max Performance Ford V-8s on a Budget (SA69P)
How To Build Max Perf 4.6 Liter Ford Engines (SA82P)
How To Build Big-Inch Ford Small-Blocks (SA85P)
How to Rebuild the Small-Block Ford* (SA102)
How to Rebuild Big-Block Ford Engines* (SA162P)
Full-Size Fords 1955–1970 (SA176P)
How to Build Max-Performance Ford FE Engines (SA183)
How to Restore Your Mustang 1964 1/2–1973 (SA165)
How to Build Ford RestoMod Street Machines (SA101P)
Building 4.6/5.4L Ford Horsepower on the Dyno (SA115P)
How to Rebuild 4.6/5.4-Liter Ford Engines* (SA155P)
Building High-Performance Fox-Body Mustangs on a Budget (SA75P)
How to Build Supercharged & Turbocharged Small-Block Fords (SA95P)
How to Rebuild & Modify Ford C4 & C6 Automatic Transmissions (SA227)
How to Rebuild Ford Power Stroke Diesel (SA213)

GENERAL MOTORS

GM Automatic Overdrive Transmission Builder's and Swapper's Guide (SA140)
How to Rebuild GM LS-Series Engines* (SA147)
How to Swap GM LS-Series Engines Into Almost Anything (SA156)
How to Supercharge & Turbocharge GM LS-Series Engines (SA180)
How to Build Big-Inch GM LS-Series Engines (SA203)
How to Rebuild & Modify GM Turbo 400 Transmissions* (SA186)
How to Build GM Pro-Touring Street Machines (SA81P)

MOPAR

How to Rebuild the Big-Block Mopar* (SA197)
How to Rebuild the Small-Block Mopar* (SA143P)
How to Build Max-Performance Hemi Engines (SA164P)
How To Build Max-Performance Mopar Big Blocks (SA171P)
Mopar B-Body Performance Upgrades 1962-1979 (SA191)
How to Build Big-Inch Mopar Small-Blocks (SA104P)
High-Performance New Hemi Builder's Guide 2003-Present (SA132P)

OLDSMOBILE/ PONTIAC/ BUICK

How to Build Max-Performance Oldsmobile V-8s (SA172P)
How To Build Max-Perf Pontiac V-8s (SA78)
How to Rebuild Pontiac V-8s* (SA200)
How to Build Max-Performance Buick Engines (SA146P)

SPORT COMPACTS

Honda Engine Swaps (SA93P)
High-Performance Subaru Builder's Guide (SA141)
How to Build Max-Performance Mitsubishi 4G63t Engines (SA148P)
How to Rebuild Honda B-Series Engines* (SA154)
The New Mini Performance Handbook (SA182P)
High Performance Dodge Neon Builder's Handbook (SA100P)
High-Performance Honda Builder's Handbook Volume 1 (SA49P)
How to Build Cobra Kit Cars + Buying Used (SA202)

*Workbench® Series books feature step-by-step instruction with hundreds of color photos for stock rebuilds and automotive repair.

ENGINE

Engine Blueprinting (SA21)
Automotive Diagnostic Systems: Understanding OBD-I & OBD II (SA174)
Competition Engine Building (SA214)

INDUCTION & IGNITION

Super Tuning & Modifying Holley Carburetors (SA08)
Street Supercharging, A Complete Guide to (SA17)
How To Build High-Performance Ignition Systems (SA79P)
How to Build and Modify Rochester Quadrajet Carburetors (SA113)
Turbo: Real World High-Performance Turbocharger Systems (SA123)
How to Rebuild & Modify Carter/Edelbrock Carbs (SA130P)
Engine Management: Advanced Tuning (SA135)
Designing & Tuning High-Performance Fuel Injection Systems (SA161)
Demon Carburetion (SA68P)

DRIVING

How to Drift: The Art of Oversteer (SA118P)
How to Drag Race (SA136P)
How to Autocross (SA158P)
How to Hook and Launch (SA195)

HIGH-PERFORMANCE & RESTORATION HOW-TO

How To Install and Tune Nitrous Oxide Systems (SA194)
David Vizard's How to Build Horsepower (SA24)
How to Rebuild & Modify High-Performance Manual Transmissions* (SA103)
High-Performance Jeep Cherokee XJ Builder's Guide 1984–2001 (SA109P)
How to Paint Your Car on a Budget (SA117)
High Performance Brake Systems (SA126P)
High Performance Diesel Builder's Guide (SA129P)
4x4 Suspension Handbook (SA137)
Automotive Welding: A Practical Guide* (SA159)
Automotive Wiring and Electrical Systems* (SA160)
Design & Install In-Car Entertainment Systems (SA163P)
Automotive Bodywork & Rust Repair* (SA166)
High-Performance Differentials, Axles, & Drivelines (SA170)
How to Make Your Muscle Car Handle (SA175)
Rebuilding Any Automotive Engine: Step-by-Step Videobook (SA179)
Builder's Guide to Hot Rod Chassis & Suspension (SA185)
How To Rebuild & Modify GM Turbo 400 Transmissions* (SA186)
How to Build Altered Wheelbase Cars (SA189P)
How to Build Period Correct Hot Rods (SA192)
Automotive Sheet Metal Forming & Fabrication (SA196)
Performance Automotive Engine Math (SA204)
How to Design, Build & Equip Your Automotive Workshop on a Budget (SA207)
Automotive Electrical Performance Projects (SA209)
How to Port & Flow Test Cylinder Heads (SA215)
High Performance Jeep Wrangler TJ Builder's Guide: 1997-2006 (SA120P)
Dyno Testing & Tuning (SA138P)
How to Rebuild Any Automotive Engine (SA151P)
Muscle Car Interior Restoration Guide (SA167P)
How to Build Horsepower - Volume 2 (SA52P)
Advanced Automotive Welding (SA235)
How to Restore Your Corvette (SA223)
How to Restore Your Pontiac GTO (SA218)

HISTORIES & PERSONALITIES

Yenko (CT485)
Lost Hot Rods (CT487)
Lost Hot Rods II (CT506)
Grumpy's Toys (CT489)
America's Coolest Station Wagons (CT493)
Super Stock — A paperback version of a classic best seller. (CT495)
Rusty Pickups: American Workhorses Put to Pasture (CT496)
Jerry Heasley's Rare Finds — Great collection of Heasley's best finds. (CT497)
Jerry Heasley's Rare Finds: Mustangs & Fords (CT509)
Street Sleepers: The Art of the Deceptively Fast Car (CT498)
Rat Rods: Rodding's Imperfect Stepchildren (CT486)
East vs. West Showdown: Rods, Customs Rails (CT501)
Junior Stock: Stock Class Drag Racing 1964–1971 (CT505)
Definitive Shelby Mustang Guide 1965–1970, The (CT507)
Hurst Equipped (CT490)

CarTech®, Inc. 39966 Grand Ave., North Branch, MN 55056. Ph: 800-551-4754 or 651-277-1200 • Fax: 651-277-1203
Brooklands Books Ltd., PO Box 146 Cobham, Surrey KT11 1LG, England. Ph: 01932 865051 • Fax 01932 868803
Brooklands Books Aus., 3/37-39 Green Street, Banksmeadow, NSW 2019, Australia. Ph: 2 9695 7055 • Fax 2 9695 7355

Visit us online at
www.cartechbooks.com for more info!

Additional books that may interest you...

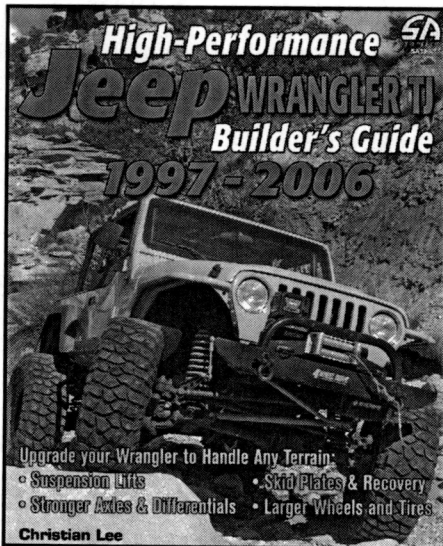

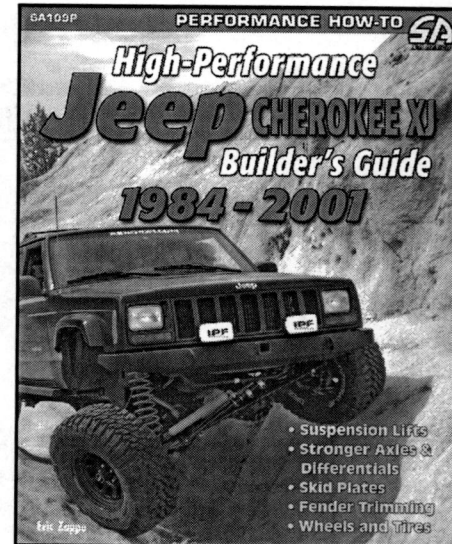

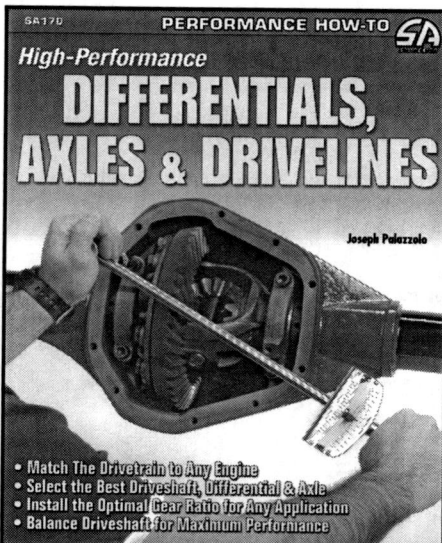

CPSIA information can be obtained
at www.ICGtesting.com
Printed in the USA
LVOW09s1146270717

542739LV00025B/178/P

9 781613 250822